Price — $12.95

Let's reach for the sun T.M.

30 original solar and earth sheltered home designs

Revised Edition

by George Reynoldson and the Space/Time gang.

Time Designs inc.

ABOUT THE AUTHOR

You're driving around the suburbs of Seattle with a nice-guy real estate agent, looking for a pleasant neighborhood and a kind of interesting house — maybe something with a few skylights or a cathedral ceiling; maybe even, if you're lucky, a stained glass window in the entry. The neighborhood you're in now looks pretty good — neat, not too far out, some variety. You turn onto a short cul-de-sac and a plexiglass bubble looms out of a clump of plants growing in the middle of the road. A weathered sign reads "The Last Extension." All of a sudden you start to twitch. What *is* this?? One house after another, like you've never seen, never even *imagined* seeing, before. Funky, freaky, festive, fantastic! Each one entirely different: one of natural wood with eaves stained in strips of orange and blue, one with a round window framed in purple plexiglass, one that looks like a house for Hansel and Gretel. All identifiably houses, but just different enough from the houses you're familiar with to give you the eerie feeling that you've been caught in a time warp, or a space warp, or a crazy dream with its own inner logic. Your mouth still hanging open, you reach the end of the street and explode into laughter. A cheerful little sign informs you, in that well-remembered writing, "That's all, folks!"

When you drive away, still feeling alternately dazed and delighted, you realize you will probably never see this place again, these houses again. Because when you try to find your way back, there will just be that ordinary neighborhood; no cul-de-sac will lead you back to Space/Time. It's in a dimension you'll never find again, a dream you'll never quite be able to remember.

Cheer up folks! George Reynoldson is alive and well and living in Sedona, Arizona. You can find him nearly any day, rocking back with laughter in his old red swivel chair at some crazy new idea. He really does dream up all these things: a ceiling covered with — can they really be muffin tins? a sculptural stair railing that proves on closer inspection to be a rusty sidewalk grating, a smoothly arched doorway of laminated oak, a tangle of beams soaring up behind an unsuspectingly innocent front door.

His background doesn't seem to explain anything about where he really came from — son of a determined Swedish immigrant, talented ex-multi-musician, University of Washington student first in music and then in building construction, for three years a 25-hour-a-day estimator then construction manager for a couple of prestigious building firms, and then in 1973, on his own, president of Space/Time. An unorthodox builder by conventional standards; never building the same house twice (he tries, but can't do it, always comes up with some astonishing new variations,) *almost* always finishing houses on schedule, being incorrigibly honest, handcrafting detail instead of buying the latest gimmick and tacking it on, generating new design ideas at the rate of roughly one every 3.7 minutes, and *still* working a 25-hour day.

All these idiocyncrasies and almost every spec house is sold in the framing stage or before. As for custom houses, he's asked to do about twice as many as he can actually handle.

He started getting serious about solar a year and a half ago, spent weeks and traveled thousands of miles around the country visiting solar equipment manufacturers and existing solar houses. The logic became more and more inescapable, and finally the challenge of designing functional *and* beautiful houses under the new constraints of solar was too much to resist. *Let's Reach for the Sun* was born.

So that's George Reynoldson, but isn't really George Reynoldson at all. The houses, for most of us, are the only permanent thing to catch hold of, and they live and breathe on their own. The owners all feel the magic. They are the first to admit their houses aren't perfect — but they're special, different, alive. There is an unavoidable realization that *people built these houses.* They are houses on a human scale, with human values dictating the design. Houses which pay careful and sensitive attention to our need to be recognized as individuals, to the knowledge within each of us which says, "I am unique."

The houses presented here are a logical extension of this recognition. They are joyful and fun-filled alternatives to split-level living, intended for families, young and young-thinking, who are not afraid of change, who are not afraid to be themselves. Most of all, they are for people who care — for themselves, for their children, for the earth.

Jeanne Erdahl

CONTENTS

INTRODUCTION

With all the current books about alternative energy sources, why do we need another? This book, like most others on the market, is a step toward helping the solar home industry happen, but that is where any similarity ends. It contains thirty original solar designs, the plans for which are available for purchase at reasonable prices. We have attempted to integrate a wide range of people's personal tastes with energy conservation and the exciting geometrics that solar design inspires.

Until now, in order to build a solar home, you had to hire an architect, a builder, solar equipment consultants, and installers. With architect fees in the range of 8% to 14% of the construction cost of a new home and contractor fees in the range of 10% to 20% of the cost, little money has been left for solar hardware and professional consulting. This book attempts to provide solutions to such cost problems by providing designs that most builders can efficiently construct. It can even give the skilled consumer an owner-built alternative.

Space/Time Designs, Inc. originally became interested in solar home design because the geometric three dimensional requirements of solar seemed to blend naturally with the type of design procedures we were already using. Homes integrating solar systems can be technologically interesting, but equally important, they can *also* be aesthetically and emotionally stimulating. People's feelings are a key ingredient in these homes and since there are as many different feelings in design as there are people, these thirty homes are just a glimpse into a field of design that is still in its infancy. These homes are meant to encompass the broadest possible range of people's feelings.

With the help of our plans, specifications, and control format, and with instructions on how to use them, most builders will be able to build a solar home with their bank's blessing. Furthermore, most of these plans can be built with conventional heating systems if the immediate expense of solar hardware is a problem. Such hardware can be added at a later date without major alterations to the structure.

Although the working drawings and specifications for the plans presented in this book will be delivered to you in a complete state, it remains for the builder to fine-tune them. Such things as structural plan checking, careful site selection, solar engineering work, proper siting, and interior decorating are best left to local professionals, and you should budget for their services accordingly. The earth sheltered designs shown in this book, such as Spacedome II and Earthworm, require that the earth supported structural elements be sized and supervised by a licensed local structural engineer. This will constitute another cost which should be taken into account. A wide range of other potential modifications is left to your imagination. Please feel free to make them if they enhance the home's liveability, solar function, or spatial enlightenment. In any event, please make any changes in a spirit of harmonic sensitivity to the home's future occupants, its site and its surroundings.

This book is primarily a plan book, and as such cannot possibly provide all the background information on residential earth sheltered design and solar technology which you should have to participate fully in your building decisions. You should read as much as possible about these rapidly evolving fields. The following books and periodicals are a suggested beginning reading list, but they should not be considered your sole sources of information.

Bruce Anderson. *Solar Energy — Fundamentals in Building Design.* New York: McGraw-Hill Book Company, 1977.

Bruce Anderson. *The Solar Home Book — Heating, Cooling and Designing with the Sun.* Harrisville, New Hampshire: Cheshire Books, 1976.

Stu Campbell. *The Underground House Book.* Charlotte, Vermont: Garden Way Publishing Co., 1980.

Rick Fisher/Bill Yanda. *The Food and Heat Producing Solar Greenhouse — Design, Construction, Operation.* Santa Fe, New Mexico: John Muir Publications, 1976.

Jim Leckie/Gil Masters/Harry Whitehouse/Lily Young. *Other Homes and Garbage — Designs for Self-Sufficient Living.* San Francisco, California: Sierra Club Books, 1975.

Edward Mazria. *The Passive Solar Energy Book.* Emmaus, Pennsylvania: Rodale Press, 1979.

James C. McCullogh. *The Solar Greenhouse Book.* Emmaus, Pennsylvania: Rodale Press, 1978.

National Solar Heating and Cooling Information Center, *Pas-*

sive Design Ideas for the Energy Conscious Builder. 800-523-2929.

Robert L. Roy. *Underground Houses: How to Build a Low Cost Home.* New York: Sterling Publishing Co., 1979.

Norma Skurka/Jon Naar. *Design for a Limited Planet — Living with Natural Energy.* New York: Ballantine Books, 1976.

University of Minnesota. *Earth Sheltered Housing Design.* Minneapolis, Minnesota: The University of Minnesota, 1978.

Alex Wade/Neal Ewenstein. *30 Energy-Efficient Houses. . . You Can Build.* Emmaus, Pennsylvania: Rodale Press, 1977.

Louis Wampler. *Underground Homes.* Gretna, Louisiana: Pelican Publishing Co., 1978.

Donald Watson. *Designing and Building a Solar House — Your Place in the Sun.* Charlotte, Vermont: Garden Way Publishing Co., 1977.

Malcolm Wells and Irwin Spetgang. *How to Buy Solar Heating without Getting Burned.* Emmaus, Pennsylvania: Rodale Press, 1978.

David Wright. *Natural Solar Architecture — A Passive Primer.* New York: Van Nostrand Reinhold Company, 1978.

Periodicals

Solar Age, published monthly by SolarVision, Inc., Church Hill, Harrisville, New Hampshire 03450.

Solar Engineering Magazine, published monthly by Solar Engineering Publishers, Inc., 8435 Stemmons Freeway, Suite 880, Dallas, Texas 75247.

Underground Space, published bi-monthly by Pergamon Press Inc., Fairview Park, Elmsford, New York 10523.

When you decide to build a solar or earth sheltered home, chances are you will be embarking on one of the most rewarding experiences of your life. Not only will you enjoy the pride of home ownership, but you will be controlling your own future energy source. In addition to the security of ownership, you can't help but feel good about doing your share to conserve our dwindling fossil fuel resources. Your concern will help preserve our valued standard of living for future generations.

During the preparation of this book, the unique requirements of thermally efficient building constantly presented us with exciting new challenges in designing logical and dynamic arrangements of planes, forms and textures. We think these homes are both functional and beautiful. In any event, we hope you will find as much pleasure in this book as we have derived from its creation.

THE ENERGY PROBLEM – SOLUTIONS ANYONE?

It all seems sort of funny, doesn't it? Solar energy application suddenly seems like a brand new idea. Yet we've been using it all our lives. Without the sun there would be no life on earth. We use it to grow plants. . .we feel better with summer tans. . .we love the colors it makes as it rises and sets . . .we've worshipped it throughout history. . .countless songs have been written about it. But we are just now beginning to think and talk seriously about capturing its power and really *using* it.

Roughly every three days enough equivalent solar energy strikes the earth to more than replace all known energy reserves. All we have to do is begin using it to our advantage. With a four-billion-year life, the sun can provide us with 3,999,999,970 years of free energy. With our mortgage banker's help, we will only have to pay for the first thirty years.

Are we really running out of gas? We have all become so familiar with the statistics concerning the imminent exhaustion of our fossil fuels that we no longer react to the facts: thirty to fifty years of oil left, a few hundred years of coal and nuclear fuels. Unfortunately, ignoring the facts doesn't make them go away. Many construction components are materials which are high consumers of energy. Plastics are a by-product of oil. The prices of aluminum, cement, steel, gypsum, and glass all rise rapidly with inflating energy costs. Doesn't it make sense for us to use these increasingly costly building materials for dual purposes — shelter *and* heat — instead of for shelter alone? Since the nitrogen for fertilizer is derived from natural gas, doesn't it make better sense to save this dwindling resource for long range food production?

The next generation (if not this one) will surely witness vast shortages of the oil and natural gas so badly needed to fuel the world's growing lust for energy-consuming goods and services. Future generations will, no doubt, witness the rapid depletion of coal and nuclear fuels even if their depressing environmental risks can be avoided through technology. The only non-depletable energy sources available to us today are those which come directly or indirectly from the sun.

Although all our current energy sources ultimately were once derived from the sun, we must now begin to tap solar energy directly. In nearly all areas of the country the amount of solar energy falling annually upon a residence *far exceeds* the heating requirements of that home. For the last two generations, however, we have been building housing as if we could avoid the laws of physics. Gadgetry at the flick of a switch, and automatically controlled devices blast away at heating and cooling our homes. We challenge nature and scoff at intelligent approaches to environmentally sound siting arrangements. We cling to grid surveys for platting and ignore the beauty inherent in sensitive building orientation. Ultimately, we should be living in a kind of human terrarium.

Several effective active space heating and water pre-heating systems are already available and countless other manufacturers are readying their hardware for market. The use of direct solar energy is here to stay. Hardware is easily available. Long-range financing is partially secured. Meanwhile, energy costs continue to escalate, spawning government tax incentives for conservation. The exciting potential of new building forms and spaces inspired by the technical needs of solar building are freely creating a new aesthetic stimulant to an otherwise increasingly mundane building industry. It is up to us to take the initiative to stop building energy-inefficient homes and start making solar heated homes more desirable by making them more beautiful.

Current earth sheltering and solar technology can help save the resources we have left. Solar is free, it's clean, it's direct. Let's reach for the sun.

PEOPLE, HOMES & NATURAL THERMAL DESIGN

Is all this for real?

It is becoming apparent that even though countless projects exist which demonstrate the feasibility of solar space and water heating, the American public remains mostly in doubt. Builders, with their justifiable fear of anything but sure-bet investments, may be contributing to this atmosphere of doubt. The fact is, however, that the technology exists. I have seen working models of effective solar homes in states from coast to coast and just about everywhere in between. In many areas, using life-cycle costing techniques, even the economics are justifiable. Without indirect government subsidies of oil and natural gas, solar energy would likely be competitive with more traditional energy sources today.

Currently, thousands of solar heated houses are either being built or are in the planning stages. Hardware production is gearing up for this demand. As reported in *Solar Energy Intelligence Report*, March 27, 1978, "Solar collector sales in the United States continued the rapid rate of growth that the industry has been enjoying since records were first kept in 1974, according to the latest survey prepared by Department of Energy. Sixth semiannual survey of 'Solar Collector Manufacturing Activity and Applications,' still in preparation, shows that 186 companies produced or imported 1,884,840 sq. ft. of medium-temperature or special (non-flat plate) collectors during the first six months of 1977. This is an increase of 54% over the previous six-month period and 168% over the amount for one year earlier. In addition, 15 manufacturers produced 3,222,208 sq. ft. of low-temperature collectors, an increase of 40% from the second half of 1976 and 105% above the amount for the previous year."

Then what is holding us back? The most likely thing is a lack of a comfortable understanding of how it all can work. As builders, designers, engineers, lending institutions and homeowners we must, in the few years remaining before a "real" energy crisis develops, learn to handle solar energy in a competent and controlled manner.

Bruce Anderson, in his book, *Solar Energy Fundamentals of Building Design* states, "In order to adopt and expand solar energy use, several factors must exist: Evidence that a change is technologically and economically practical; a manufacturing industry to produce components; a service industry to maintain the components; designers to incorporate them into the design of buildings; and contractors and clients and users willing to adopt this new change."[1]

The first three of these factors appear well on the way to realization. You, the reader of this book, are the crucial remaining element — the individual *willing to adopt these changes.*

So what's so different about a natural home . . .or is nature just plain beautiful?

Except for additional security, comfort, and exciting inner spaces, solar home living is probably no different from your present style of living. You'll bring the same crazy, unique family habits into your new home. You'll still find peas under the table and dirty socks under the bed. Energy awareness may be a big side benefit, but generally the use of solar energy changes heat bills, not people.

The technology of solar building is new, but the functional and aesthetic design considerations are similar to those for conventional housing. The design of a solar heat-assisted or earth tempered home involves a complex hierarchy of trade-offs and considerations. Factors such as cost, form, ease of construction, insulation, ventilation, site and room orientation, room arrangement, aesthetically stimulating exteriors and family livability result in a set of sometimes diametrically opposed design parameters. In both natural and conventional design, there is the need to constantly compromise and balance all facets of design for all building uses. But the requirements of solar are introducing some new variables.

Evolution in building forms has always taken place as new technologies reveal new possibilities. Thus, the geometrics required by solar technology can give designers a new rationale for form, and the simplicity of an effective solar design can create a new perspective of visual purity. The inner spaces of a light, warm greenhouse, the muraling effect of a collector wall, and the contrasts in form and texture that are inherent in solar energy collection and storage equipment fulfill our constant need for aesthetic stimulation and enhance our daily surroundings. With natural design, form doesn't only follow function, form becomes function. Its very simplicity means buildability.

While we are concerned on one hand with the correlations between design and effective solar function, we are equally concerned with the relationship between design and the needs of people for emotionally pleasurable and enlightening surroundings. If you are considering building any new home you have, no doubt, thought about what you are seeking in terms of number of bedrooms and bathrooms, spatial relationships among areas of your house, and so on. More intangible, however, is how you want your home to *feel.* Should it be light and airy; warm and cozy; clean, sleek and geometric; or filled with joyous clutter? You know that your feelings about your home, site, and furnishings affect your daily affairs. You may not have considered the logical extension of this fact: your surroundings can change you. They can delight you or they can stifle you. Mundane surroundings lead to mundane lifestyles. Traditional surroundings evoke traditional lifestyles. Funky designs, using such things as barnwood, stained glass, and old weathered sailing masts evoke a laid back lifestyle. Contemporary styles using such materials as chrome, glass, polished oak, and geometric forms evoke high energy living. Man, an emotional creature, finds stimulation and a sense of self through his surroundings. In the increasingly mass-marketed world of the tract house, it is vital that you identify and respect your individual emotional needs.

What is a naturally heated home?

Most people think solar energy is new in house design. The fact is that it has been used throughout history to heat

1. Bruce Anderson, *Solar Energy Fundamentals of Building Design,* New York: McGraw-Hill Book Company, 1977, p. 26.

and cool our buildings. The American Indians in Mesa Verde, Colorado, and the Acoma Pueblo in New Mexico used it to passively heat their homes during winter and cool them during the summer. The Hopi's hogan and the Eskimo's igloo are other early examples of building design for thermal comfort. Native Americans traditionally built to dampen the diurnal temperature swings through complementary use of solar and earth sheltering. More broadly, sun, earth, and building form have *always* been part of human environmental planning, until just the recent past when, through technological advances, energy suddenly became cheap and plentiful. But now our profligacy is forcing us to return to a more responsible attitude towards our resources. And this time we can combine with age-old energy wisdom our technical knowledge to come up with newly efficient and meaningful ways to use sun, earth, and form.

In order to apply solar heating to our homes, we must provide for a man-made means for *collection* of the sun's rays, *circulation* of the heat and *storage* of this heat until it is needed for indirect circulation. *Earth sheltering* is an additional technique which can be used concurrently with any type of solar heating to cut down the overall heating or cooling requirements by minimizing daily and seasonal temperature fluctuations. A more complete discussion of earth sheltering follows this section.

There are two commonly applied ways to accomplish solar space heating. The first is known as *active solar heat assistance.* Collection occurs when sunlight strikes a single or double glazed panel backed by a non-reflective absorber plate. The fluid (usually air or water) surrounding this plate (or enclosed within it) is heated to a usable temperature. It is then "actively" transported by mechanical means (pumps, blowers, pipes, ducts, etc.) to an enclosed storage material with mass equal to or greater than that of the carrier fluid. As the heat is needed for space heating, it is drawn from the storage mass and by some type of heat exchange process enters the conditioned space in the form of warmed air.

The other method of collecting, storing and circulating solar gained heat is commonly known as *passive* or *natural solar heat assistance.* These systems use natural thermal processes with minimal mechanical assistance. Sunlight entering a window or skylight and striking a dark-colored building material is a common example of a passive solar effect.

Heat circulation occurs by natural convection, conduction or radiation. Storage is the direct result of incorporating dense building products with high heat capacity between the source of heat and its point of consumption, thereby utilizing the time-lag concept associated with high mass storage. When the sun goes down, these materials slowly dissipate their heat to the conditioned living space.

Depending on the quantity of storage mass, size of the solar aperture, amount of building heat loss and the living habits of the building occupants, some house spaces can remain comfortable for many hours and possibly even days without back-up heat assistance.

Many of the designs prepared for this book are commonly known as *hybrid solar systems* and utilize both active and passive solar heat assistance methods.

The rest of this section will briefly discuss the design principles which we have attempted to apply in presenting the following thirty plans. To analyze these principles in detail is beyond the intended scope of this publication. The books listed in the suggested reading list and bibliography offer many descriptive and entertaining examples of other solar buildings.

In most cases, cost of construction is carefully balanced against aesthetics and solar effectiveness. Fortunately, the needs of a good solar building are in many ways the same as the needs of any beautiful and cost-effective non-solar structure. Using cost-effective principles such as minimizing gross spatial volume, perimeter wall area, and square footage also work for minimizing total heat loss. The simple building forms which solar collection often dictates can in themselves lead to efficient construction systems. The implied length-to-depth ratios along the sun wall axis dictate reasonably efficient structural simplicity. In fact, the two earth sheltered domes (pages 69 and 129) shown in these designs are the theoretically perfect form for minimizing heat loss and maximizing structural integrity. Additionally, they approach the most commonly seen shape in the cosmos — the sphere — for those of you concerned about the metaphysical properties of form. Although building costs always necessitate some spatial compromise, we have ensured that solar technology has not caused a compromise of beauty or livability in any of the homes in this book.

In designing a solar home, energy conservation measures must come first. All of the designs in this book include 2x6

24 in. o.c. exterior walls for maximum insulation. These designs include such features as a minimum number of north facing windows, maximum roof and crawlspace (or slab) insulation, "tight" building details, entry air locks, garage to the north side of the home, steep north roof slopes, partial north earth sheltering and slightly elongated southern plane exposures. These features will be maximized when put in combination with external energy saving measures such as north earth berms, south oriented deciduous trees, reflective yard surfaces, etc.

An especially important conservation necessity is to understand the relative U-values or R-values of building products, some of which you will have to specify from locally available building supplies. A U-value is a standard measure of the ability of a material to *conduct heat* through a surface. It can be defined as the rate of heat loss in BTUs/hour through one square foot of building surface when the indoor to outdoor temperature difference is 1°F. The R-value is the reciprocal of this number and is a measure of a specific material's total resistance to heat flow. For example, 1 inch of polyurethane insulation has an R-value of about 5.88 while 1/8 inch of glass has an R-value of about .89. By adding all the R-values of a particular wall or roof cross section, one can methodically calculate the ability of any building surface to hold a constant temperature. By comparing R-values to the price per square foot of the material, the energy cost effectiveness of all building materials can be evaluated.

Nearly all these homes are designed with both passive and active solar features. Current debate about the advantages of either system is beyond this discussion. It should be pointed out, however, that it is common among architects, engineers, designers, and users of solar energy projects to be biased toward what they are most familiar with. That usually is their last best project. It is a challenge to everyone planning solar projects to be open to *all* workable design solutions.

Passive features incorporated in these designs include heavy mass walls, both vertical and tilted glass greenhouses, and direct gain floor slabs. Let's take a closer look, now, at how both passive and active systems have been integrated into these thirty designs.

Several of our plans are designed with *heavy mass glazed walls.* As used in test structures at M.I.T. and as worked with in France by architect Jacques Michel and Professor Felix Trombe, these walls can provide considerable heat to the conditioned living space of a home. As the sun strikes the air between the glazing and black mass wall (which can be built of a variety of dense materials such as concrete, brick, stacked eutectic salt containers, or stacked water-filled containers), this air is warmed and begins to rise. Ducts or vents at the top of the wall allow this hot air to flow into rooms, as shown in Figure 1. The air is replenished by cool air intake vents at the bottom of the wall. At night, when no energy input exists,

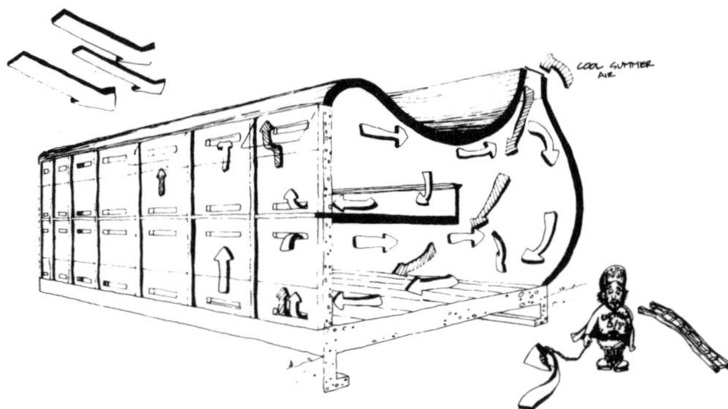

DAYTIME OPERATION OF MASS WALL

The mass wall acts as a natural collector. Convection draws warm air out openings at the top of the wall and pulls cooler air in through openings at the bottom of the wall. During the cooling season cool air is drawn from north windows, pulled through the bottom of the wall and is vented to the outside at the top.

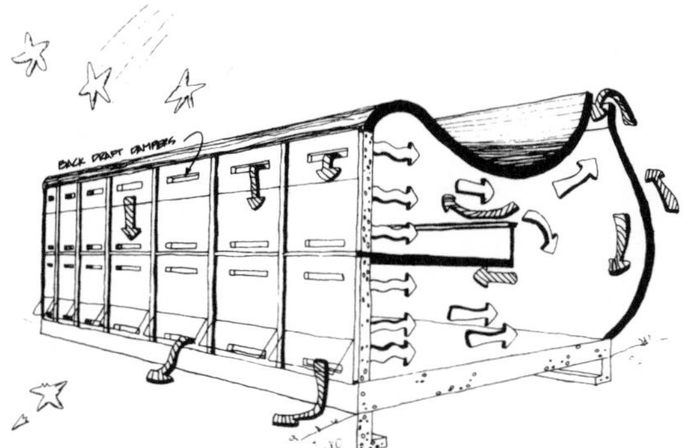

NIGHTTIME OPERATION OF MASS WALL

Stored heat radiates to the conditioned space during the heating season. During the cooling season back draft dampers at the top of the wall are left open so reverse thermosiphoning can ventilate warm air to the outside.

Figure 1

the heated mass radiates heat to the conditioned space thereby stabilizing the temperature. Potential problems of overheating and temperature stratification can be avoided with the addition of fans to distribute the warm air evenly throughout the home. The "summer fan" mode in most conventional forced-air furnaces, provided with sufficient cold air returns near the heat source, can accomplish this if sized carefully. Provision for some additional storage mass along the path of the return air can be added to absorb excess heat and save some for night-time back-up heating.

During summer months, vents at the top of the wall can be opened to exhaust warm air out of the structure. With north windows open, the wall's natural thermosiphoning characteristics will provide ventilation. Openings in the wall, vent sizes, and wall thickness must be carefully worked out with your engineer. The effectiveness of a mass wall is greatly increased if some type of movable insulation is placed on the

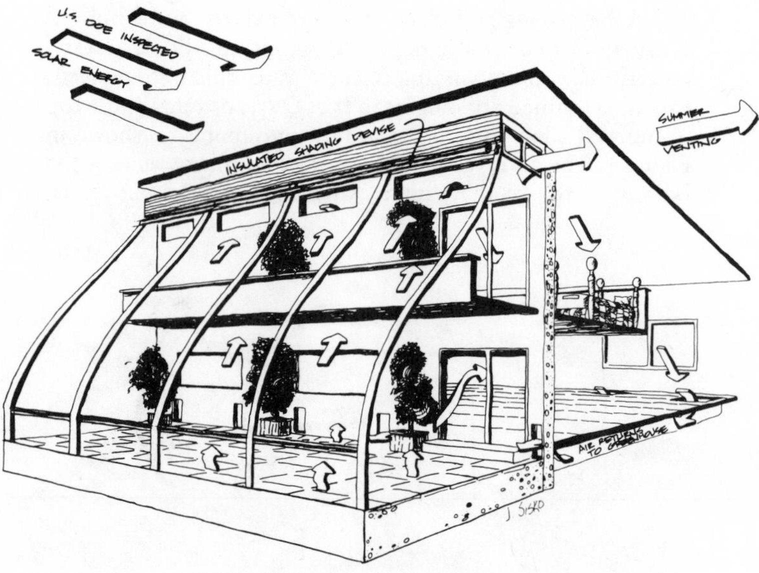

THE PASSIVE SOLAR GREENHOUSE

Solar energy warms the greenhouse and temperatures are stabilized by dense mass objects. Heat enters room space by convective air flow through controllable openings or by radiating off the north greenhouse wall.

Figure 2

outside of the glazing, or between the wall and the glazing at night or during cloudy days. Such hardware may soon be commercially and inexpensively available. Aesthetically, a mass wall can be used to surprising advantage in your design. Extraordinary lighting effects can be achieved. Try a patchwork of leaded stained glass or try stacking deep-toned colored water containers.

The second common passive concept utilized in these drawings is the *solar greenhouse* in Figure 2. A good reference source on this subject is *The Food and Heat Producing Solar Greenhouse — Design, Construction & Operation,* by Rick Fisher and Bill Yanda.

Basically, solar energy enters the greenhouse in the form of short wave length radiation. As shown in the diagram, these light waves strike and warm absorbent objects. Materials such as a black concrete slab floor, or dark colored water containers are commonly used. Heat is emitted from these objects in the form of long waves, which are not readily returned through glazed surfaces. This principle is commonly known as the "greenhouse effect." Heat is provided to the conditioned space through vents from the greenhouse. Here again, movable insulation over the glazing can greatly increase the heat assistance potential of this type of project. Two great side effects of this type of solar heating are the aesthetic potential of the greenhouse and the food and income which greenhouses can produce. To make a solar greenhouse successful, it helps to know something about gardening, but if you don't, maybe you could rent that portion of your house out to some industrious budding horticulturalist living on your block.

Greenhouse-type designs should work best in areas such as the Pacific Northwest where a high percent of solar energy is in the form of diffused sunlight, since usable temperatures with greenhouse-type designs are much lower than with most active systems.

Another passive solar heat assistance concept used in this plan book and illustrated in Figure 3, is commonly known as *direct gain passive heat assistance.* Any south facing window is an example of this. (See, you already *do* own a solar home.) To prevent daytime overheating high mass materials are incorporated into the home. Heat slowly builds up deep within the fabric of the structure until the sun goes down and then this process reverses. Insulating shutters, louvers, or shades need to be fixed over the glazing to prevent most of the heat from escaping to the outside when the sun isn't shining. Since a single pane window loses heat about 15 to 20 times faster than an insulated wall, the total R-factor of such insulating devices must be carefully considered. They should be built tightly so that air can't infiltrate the area between the glass and the shade. They should also be secured at both the top and bottom to prevent reverse thermosiphoning of air from the room.

One home in this plan book incorporates all of the passive systems we have just discussed. In "The Greenhouse Gallery," we have designed a glazed mass wall on the north wall of a greenhouse, as well as some direct gain areas. The design should result in uniformly controlled heat.

One unique application of passive solar design, the double shell design, or "house within a house," was developed by San Francisco architect Lee Porter Butler. (See Plans S/T-31, S/T-34 and S/T-35). In a conventional house, the walls, floors and ceilings are exposed directly to the extremes of the outside climate. In the double shell house, the gravity geo-thermal envelope is created by separating the interior surfaces from the exterior surfaces with a protective blanket of constantly circulating air. The interior walls and other surfaces are thereby isolated from the extremes of the outside climate. In the winter, heat enters through south-facing glass in the greenhouse and rises to the highest area of the structure. As the air cools, the force of gravity pulls it down to the lowest point in the system, the crawl space, and causes a predominately counterclockwise force on the air envelope. The heat gain is absorbed by the structure and associated thermal mass, and the constant circulation of air distributes the heat gain uniformly throughout the building. At night, the air movement largely reverses, and the ground mass, which has been warmed during the day by the solar heated air, replaces the majority of the heat loss whenever the air temperature falls below the mass temperature. In summertime, when the house is receiving more solar heat than required to maintain comfortable temperatures, vents at the top of the structure (see S/T-35) open to draw in earth cooled air through underground cooling tubes or through direct outside openings into the crawl space. Impressively, the double shell house will maintain comfortable interior temperature 90% of the time, even in northern climates.

The scope of other passive design ideas you may wish to incorporate within our plans is limited only by your knowledge of thermal processes and your imagination. Don't be afraid to experiment beyond our plans. Who knows, you may discover the better mousetrap the world is looking for. Be sure to think ideas out thoroughly, first in the heating mode and then in the summer cooling mode.

Now, let's consider active solar systems. All of our designs have some active or mechanically-controlled devices used for solar heat assistance, though in many cases the only active part of the system is the hot water pre-heat feature. Basically, five components are needed for most of these designs: collectors, an isolated storage mass with good air-heat exchange properties, duct work or pipes, a pump or fan to move the fluid between the collectors and storage, and a back-up system

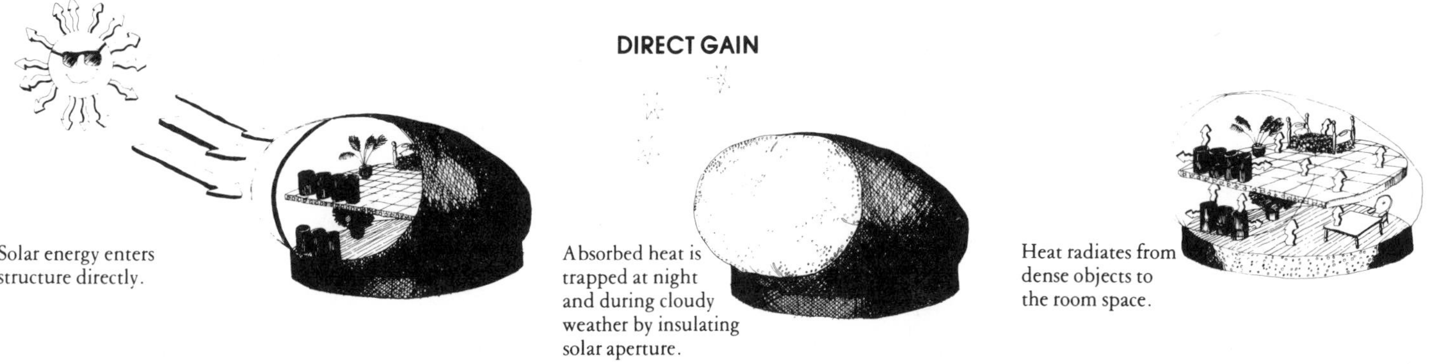

Figure 3

to distribute the heat to the conditioned space and to provide supplementary heat when necessary. A typical active air heating system is diagramed in Figure 4.

First, the collectors. Although there are literally hundreds of types of solar collectors, only flat plate type collectors will be discussed here. The two most common flat plate collectors are those where either water or air is heated by being drawn across a black metal or plastic container covered by one to three layers of glass or plastic. Although water-type collectors usually achieve better efficiency ratings, we generally have specified air-type systems because of their lower maintenance, generally lower cost and fewer potential problems associated with freezing weather. This is not to imply a bad future for water-type heating systems. They generally work best, however, when cooling is as important as heating.

An air-type collector can be built relatively easily and it is possible to build one yourself. You must be very careful to have even the simplest system designed by a professional who is very knowledgeable about mechanics, heat loss and heat transfer calculations. Once designed and detailed so that the system is a completely closed one, with no uncontrolled air gaps, building a custom collector is quite simple. Usually however, after considering the time and engineering requirements, it is easier, less expensive, and safer to purchase a well-manufactured factory-built system. You should do additional reading on collectors before you decide to invest.

For certain, know enough to question the relative effects of double and single glazing, solid versus perforated absorber plates, wood versus metal frames, black non-reflective paints versus selective coatings, and the long range durability of all materials which make up the collector. It is especially important that all collectors be built tightly to prevent air and water infiltration and loss! Be sure the answers you get to all questions are based on scientifically derived test data. Reliable hardware sales people will be proud to share their

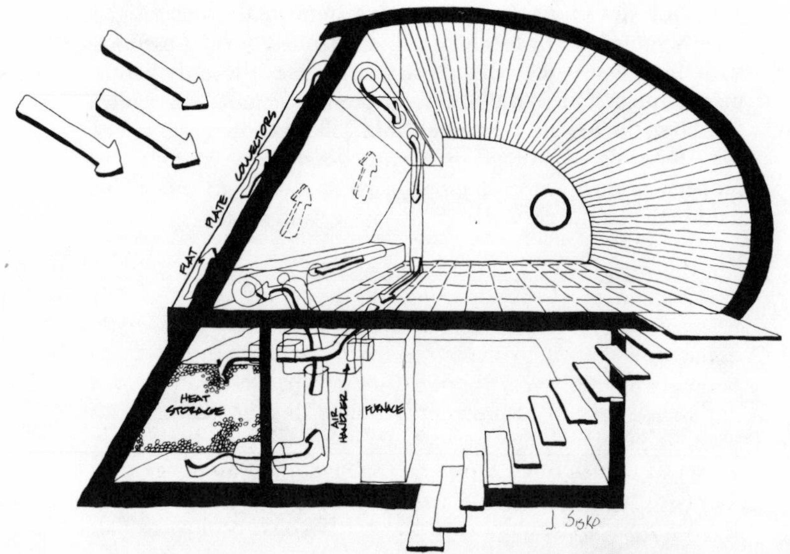

ACTIVE SOLAR COLLECTION AND STORAGE

Sunlight strikes collectors and heated air is drawn from the top of the collectors and pulled downward or diagonally through the storage medium. Heat is absorbed in the storage bin and cooler air is recirculated through the collectors.

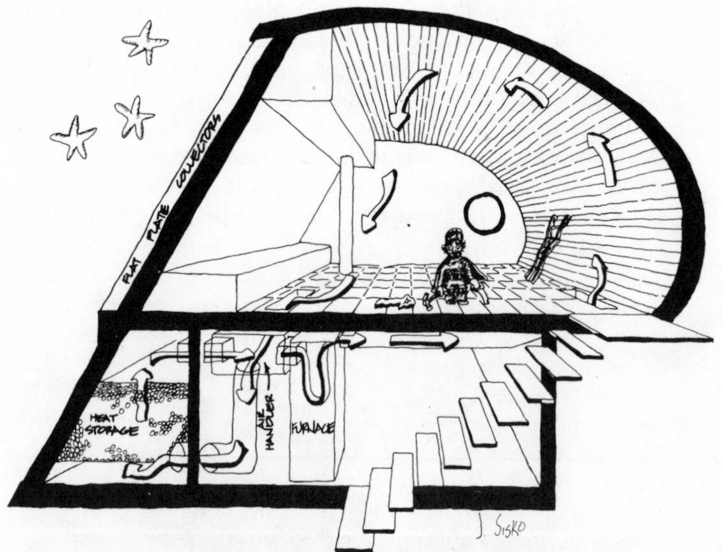

ACTIVE CIRCULATION OF STORED HEAT

Stored heat is actively circulated to rooms through the air handler or the forced air backup system. Cooler room air is returned to storage.

Figure 4

collector test results, and good marketing people will know the advantages and limitations of their equipment. There have been cases where companies have made grossly misleading claims to potential collector purchasers. Don't get caught by a "solar hustler."

The second primary component of an air collector system is a storage medium. Usually, a large rock bin, a room full of water-filled containers, or stacked eutectic salt trays are used. This latter system, still in the developmental stage, features plastic trays filled with sodium sulfate decahydrate, a material which characteristically changes phase at about 90° F. With a heat of fusion of about 9,860 BTU/cubic foot, it has the advantage of larger amounts of thermal storage in less space than water or rock. Air blown from the collectors is held in the storage mass until it is needed for space heating. The heat capacity of the storage material and the volume of air filling the voids in the storage area will greatly affect the design and sizing of your system.

A useful tool for rock bin sizing is the manual sold by Solaron Corporation[2] of Denver. The size and type of rock used can be critical because it affects the static pressure of the system and the storage area's ability to stratify properly. Other manufacturers will provide you with other design tools they feel are useful.

Various other materials can be used for storage, but they must all be carefully engineered. Having witnessed many storage experiment failures, such as leaky water containers, leaky salt trays, rocks graded so that they fail to permit satisfactory air movement, and water condensation in storage rooms (be sure to seal them well and keep above grade if possible), I suggest you stick to proven systems.

Sticking to basics applies as well to the third component of an active solar heating system: *ductwork.* Use the KISS (Keep It Simple, Stupid) system whenever possible. Without question, the more ductwork, the more pipe friction; the bigger the blowers, the more space wasted; the more strain on motors, the *less efficient the system!* Quality of workmanship is crucial. Air leaks can cause two big problems: temperature stratification within the structure and insufficient input to storage. Each duct bend can cause unnecessary pressure on the air mover components.

2. The Solaron design manual is useful for total air collector system design with rock storage. — see Bibliography.

Most plans in this book are designed to provide sufficient duct chase area for cold air and warm air risers and manifolds. Your engineer may need more or less room than we have provided. Usually these changes can be easily accommodated by furring out a closet or some other low-priority space.

The fourth component is both the brains and the brawn of the system. Commonly known as an air mover or air handler, this fan-driven blower unit with the aid of sensor probes and a control logic system, automatically tells itself when to collect. Controls and sensors can be purchased separately from the blower unit, but some new air handlers are being marketed with integral controls and directly wired sensor probes. In most cases, it is probably less expensive to buy a packaged unit than to build your own. The advertising in such publications as *Solar Engineering* magazine and *Solar Age* magazine is a good source for specific product information.

The auxiliary heating system completes the list of basic components of an active solar heat-assisted home. A conventional forced-air electric, gas, or oil furnace is probably the best because it already has a built-in air moving system which can tap storage. Its output should be sized as if the solar system didn't exist because during long periods of overcast weather it will need to assume the full heating load.

Why build underground?

In a macro-economic sense, all the techniques for energy gain, storage and conservation described so far are simply a rearrangement of the structure of the capitalization of our primary economic resource, energy. Put another way, every dollar one spends to make his living environment both as permanent and as energy non-consumptive as possible, is a dollar less debt of future energy claims that he will have to incur and still maintain his same standard of living. It is simply a way to reduce one's future debt obligations (mortgage payments excepted here) and reduce both individually and collectively our indebtedness to the earth's own ecological balance of payments.

Earth sheltering techniques provide a very simple model of this fact, mostly because of the inherent properties of strength and thermal efficiency of their most suited building components — concrete, steel, and the earth. Concrete, when steel reinforced, is the strongest practical building material known to man. Its permanence, well illustrated by still-existent ancient Roman roads, implies non-consumption of the earth's resources. Concrete structures, if properly designed and built, are relatively fire, decay and earthquake resistant, and have a possible lifespan of several hundred years with a minimum of maintenance. Furthermore, the shell itself and its waterproof skin can be further protected by a thick layer of earth, which, unlike traditional building coverings — paint, siding, shakes, stucco — does not quickly decompose in the face of rain, wind, freezing, and even the sun's rays. Underground reinforced concrete buildings, whether precast or poured in place, also provide an excellent heat sink in the structural fabric of the building itself.

If this type of design can really last so long and remain relatively maintenance-free and energy efficient, the long-range effects of being able to avoid indebting the earth's resources for construction purposes *alone,* not to mention energy savings and the avoidance of debt by the structure's inhabitants, could project a profoundly different future from the frequent reconstruction cycles dictated by today's wasteful policies of planned obsolescence. That is to say, future energy and construction real wealth can be in your and your children's hands instead of in the hands of the oil companies, lending institutions, and utility companies. *And,* if your home is totally hidden under the earth, maybe even the real estate tax assessor won't be able to find it!

Designers who are aware of the potential for integrating modern thermal and structural engineering concepts will be able to combine new and beautiful passive solar building forms with the plastic and free-form properties of concrete and steel in a way far beyond the imaginations of the ancient Roman and native American inventors. That earth sheltered building is still considered risky by many lenders only indicates their ignorance of environmental history; this is obviously not a new building principle!

The two underground dome designs presented in this book are natural examples of how to use concrete, steel, and earth in such a way as to optimize energy containment, circulation, and structural efficiency. The hemispherical dome is the theoretical optimum shape for energy conservation because it has less surface area for heat loss than any other possible building shape of the same volume. It saves on overall building materials because of its unique properties of stress/strain distribution and its surface area properties as well. The lack of efficient forming systems, and the fact that most local building departments require local engineers to calculate the stresses separately, stamp the drawings for approval, and subsequently supervise the construction, may be the only factors keeping this method of construction from gaining rapid popularity. If substantial public interest results from this first public exposure of our Spacedomes, Space/Time Designs will be developing a universal engineering and forming system.

The total possible benefits of earth sheltering buildings is beyond the scope of this limited discussion. However, besides protecting the building from naturally destructive elements (and potential manmade ones such as smog, acid rain and nuclear radiation), the principle advantage of earth sheltering is that it functions to equalize interior temperatures *seasonally* just as heat storage mass equalizes interior temperatures over the period of a few days. As the seasons change, the earth temperatures adjacent to the surface of the building remain fairly constant, so in the summer months when ambient air is 95° F., your cooling equipment relaxes as this air "sees and seeks" the 55 to 58 degree temperatures along the surface of the building. This works similarly in winter months when your solar gain system or back-up heating system only has to overcome 55 to 58 degree surface temperatures while it could be sub-zero outside.

The construction costs of underground structures need not be much if any more than conventional construction, especially as industry workers make an experience-level adjustment to the use of concrete and steel from wood stick-framed structures. With efficient forming and excavation methods and optimum natural site conditions, such homes could be even less costly to build. Waterproofing and excavation costs, for instance, would be traded off against those for siding and roofing. In the future, it seems possible that the costs of lumber, plywood, wood, aluminum and vinyl sidings, wood shakes and bituminous roofing may inflate in cost faster than concrete, steel and glass, especially in areas of the country where natural scarcity exists. Concrete and glass are usually locally manufacturable because their major components — water, aggregate, silica and limestone — are locally available throughout the country. Forestry products, aluminum, vinyls, and bituminous materials often are subject to the costs of shipping long distances which obviously make them further dependent on the fossil fuels. Their production technologies are high energy consumers as well.

Finally, by far not the least benefit of earth sheltered building design is the fact that it affects our natural environment less than any other building technique. As a result of replacing earth and natural ground cover after construction, it has as yet unimaginable potential for changing our current wasteful site planning habits on a city planning scale as well as on neighborhood and individual site planning levels. Consider, for example, the appeal of having your entire lot, *after* the construction of your home, still available as a playground for your children and a garden for your family!

The effectiveness of earth sheltering varies with soil types, ground water conditions, humidity, earth temperature depth curves and the seasonal extremities of the local climate. Before you go ahead with earth sheltered building, you should study local soils, site, and climate conditions carefully, and read as much on the subject as possible. On-site inspections of other earth sheltered buildings is also important, because you can *see* for yourself the subtle changes in building methods which earth sheltering requires. Several books are noted in the reading list to give you a background in the area.

Earth sheltered buildings need not be "dull, dark, and damp" as many unimaginative opponents to this form of natural home design maintain. On the contrary, these buildings may well be the harbingers of a rapid evolution in the design of human environments.

SOLAR – TO BUILD OR NOT TO BUILD

How to decide to build a Space/Time thermally efficient home

You should begin by studying the technical, functional and economic feasibility. To work technically, *your site must have sun* and should preferably also have winter wind protection. Talk to others in your area who have built solar houses. By all means, consult with a qualified solar consultant. To find one, try calling your local university as a start. Look in the phone book under heating systems or solar equipment manufacturers or call some mechanical engineering firms. Interview your solar expert carefully. Make certain he is as current with the dynamics of this rapidly changing industry as possible. It is important to know how much solar energy arrives at your site in the form of diffused radiation, as this may influence you toward a particular approach.

Assuming sufficient sun exposure, look at the plans in this book to find a design that is functional for your family needs and fits the topography of your site. The home you select should feel aesthetically stimulating to you. Don't be afraid to be different in house styling, but be sure your design is at least relatively acceptable to the neighbors. Check to see that it meets local architectural control considerations. Since it is likely the only solar home on your block, it can't be expected to look like the others. Most people will learn to adjust to the new beauty of solar.

Having picked a design you like and one that you know fits your site, look at its economic feasibility. Study its resale potential in the market-place. The first step is to determine cost. Many builders give ballpark guesstimates on an approximate cost-per-square-foot basis. For most of these homes an additional $2.00 to $6.00 per square foot should be allowed for solar equipment, fixtures, glazing, mass storage, and underground-related excavation, backfill and waterproofing. Passive solar designs are probably closer to $2.00 per square foot extra while active designs can be closer to the higher costs. Resale is equally important to builders and owners. As more solar homes are built and utility bills become less tolerable, the solar home market will certainly take a strong leap forward in the next few years. Talk to the most knowledgeable and reputable mortgage bankers and real estate people in your area. As most of these people remain uninformed about our energy future, energy conservation needs, and solar energy potential, you must be discriminating about whose advice you take. Chances are, if many other people in your area are thinking solar, resale opportunities look bright. If your banker says he'll lend 75% to 80% of full appraisals, you generally know your market turnover risk is perceived as normal to him.

Another deciding factor will no doubt be the availability of federal and local tax incentives. Up to a $2,200.00 federal tax rebate is now available for residential solar heating systems. A comprehensive bill crediting passive solar features is likely in the near future. Other incentives such as local real estate tax exemptions and state income tax deductions are available in many states. If you finance over thirty years, the interest is tax-deductible, whereas fuel bills are not. Considering these tax advantages, solar is not only a long-range bargain, but can also provide you with a short-range (one to seven years) positive cash flow. What a deal!

If the practical side seems sound, look at the more idealistic side of the project: the pride of securing your energy future and doing your part to conserve scarce future resources; the functional purism of a building that works as a solar collector and a solar storehouse; and the family closeness than can be enhanced by an ongoing greenhouse project. Additionally, it's just plain fun to participate in a simple technology and check collector and storage temperatures occasionally.

One final way to decide on solar is simply to rationalize. Many people will no doubt want solar just because they want it. Some people think nothing of purchasing a $15,000 automobile knowing its value will be only $3,000 to $5,000 in four years. With the same $15,000 you could guarantee yourself free energy, finance it over thirty years, and see it appreciate in value as well.

Before you begin to build your solar home there are a few more qualifications. Patience and perseverance are essential. You must be able to get along with a lot of new people, most of whom may know less about solar energy in the home than you. You must be able to sift out truth from fiction. You must demand, respect, and appreciate quality in construction. You must make it clear to all the people you work with that you will be fair and reasonable with them and thankful for their best possible workmanship. Constantly be open; show awareness that nobody has all the answers about building, much less about solar building. Candor and honesty will pay you big dividends.

When you decide to go ahead with solar, chances are you will be a pioneer builder or owner in your area. You will be bucking current wasteful trends and are to be applauded for your courage. The possibility seems strong that five years down the road you may be envied for taking the first step in your area while others face the full burden of 50% to 100% increases in energy costs, or are trying to jump onto the solar bandwagon faced with highly inflated building costs. Plan carefully, proceed with cautious trust, but don't look back. Even if you choose a design other than one of ours, good luck. Without question, you are in for a rewarding experience.

Watch out for pitfalls

Solar home building often brings out the best and worst in people simultaneously. We are admired for our innovative tendencies and at the same time criticized for our mistakes. One should not expect to build a house, much less a solar house, without making mistakes. But if you are careful, most of the major pitfalls can be avoided.

Basic to the success of any solar system is careful attention to traditional energy conservation measures. Even the best solar installation will fail if the house is incorrectly sited or insufficiently insulated and sealed. *Optimal siting and adequate insulation are probably the two most important factors in building a successful solar home.*

Underestimating the scope of the entire project is also not at all uncommon. At times the cost, complexity, and time required to complete the task at hand can seem formidable. But builders often take complexity in stride only to find out after totaling final costs that what may have looked simple actually required a lot of additional time, labor, materials, and control supervision. Building a solar home requires a new level of understanding. One *must* read about solar trends and new ideas. A minimum reading list should include the books and periodicals listed in the introduction. If you study these sources, one of the biggest pitfalls, ignorance about solar energy, should be well on the way to being eliminated.

If you are getting into active solar systems, avoid buying untested hardware and fixtures. Several reliable active space heating systems are currently being marketed. By the time you read this book, efficiency standards for collectors, both vertically oriented and tilted may be available. In order to keep current on this, you may call the National Center for Heating and Cooling at 1-800-523-2929 and find out how to get current test data.

Inaccurate sizing of active collector area and storage capacity to your climate or pocketbook is another danger. The larger the collector area you choose, the greater the reduction of your heating bill. In many sun zones, one hundred percent solar heating can be technically achieved. Economically however, this is usually impractical. The storage size must be carefully balanced with respect to the collector area and climatic conditions. You can approach this problem sensibly by consulting with a solar expert in your area who is sensitive to your requirements. Work only with solar consultants with a substantial history of dependability in solar or a related field such as mechanical engineering. Nearly all the major manufacturers of solar hardware are staffed with solar or mechanical engineers capable of sizing storage, collector area, and air ducts. It may be a good idea to have an independent professional solar consultant familiar with your area review your manufacturer's claims.

Although these designs have suggested storage and collector areas, in many cases you may want to make adjustments to your climate and budget. All reliable hardware salespeople will be sensitive to this and will help you make sensible decisions. Don't get caught by the solar hustler who thinks adding as many collectors as possible to the roof and blowing the hot air over as many rocks as you can fit into your basement will result in a maximum efficiency system. Not so! Air flow, collector area, and storage capacity must be balanced!

Beware of "cheap" active systems. There aren't any, and probably never will be. For now at least, solar is fun and promising, not inexpensive. In most cases, you must look to future years, not payback figures, to justify solar heating.

Passive systems such as greenhouses, mass walls, and direct gain floor slabs potentially have similar pitfalls. Sizing of aperture and mass, insulating apertures at night and in cloudy weather, providing shading, and calculating air flows for both heating and cooling modes are all very critical. Research being done in this area should provide us with means to better predict specific behavior of passive structures. The passive designs in this book work diagramatically, but they also must be adjusted to your climate and budget. Local solar consultants will again be of great help to you.

The risk of making mistakes in balancing passive designs is much less than with active systems. Making modifications to most passive systems after construction is simpler because of easy accessability to their components. Adding mass to a greenhouse, adding exhaust fans, bringing in additional outside air, and adding shading and insulating devices are often easily accomplished. Altering ductwork, however, or adding storage capacity in an active system is often nearly impossible in a finished house.

Earth sheltering is not without potential pitfalls either. Nightmarish disasters may occur because the owner or builder is trying to build without proper respect or knowledge of the basics of structural building principles, frost heave considerations and hydraulics, or the dissimilarity of adjacent soils or incompatible building materials. I do recommend that anyone getting into earth sheltered building hire a soils engineer, a structural engineer, and a top quality concrete contractor. You may spend an extra $1,000.00 for their services, but you may well save yourself embarrassment and redoing costs later. Don't expect the standard basement/foundation contractor to perform the quality of concrete finish and steel placement necessary to insure a structurally sound job. Read a lot and ask around to many sources about waterproofing. Make certain that all earth-supporting concrete is poured with a waterproofing admixture added to it and take core samples for structural verification. *Knowing* that your earth sheltered home is designed and built to take the seismic loads and hydrostatic pressures unique to your site and area will help you sleep better. Your banker will sleep better too!

Staying current in the latest state of the solar art will help you avoid most pitfalls. Be methodical in your approach. Allow yourself time to investigate such things as "solar rights" legislation and current tax incentives. Be reasonable about your performance expectations. Being willing to turn down your thermostat won't hurt, either. All the answers aren't available yet, but if you dedicate yourself to learning, you'll probably get enough information for your purposes and have a great time doing so.

Planning your natural Space/Time home

How should you plan a Space/Time solar home? Very carefully! As much care must be given to detail during the planning stages of the project as must be given to the building process itself. Thoroughly understanding the sequence of events can help you alleviate the fears you may have about building. Let's look now in some detail at the planning process.

The first step is to be excited about the project. Attitudes are mirrors of the mind and reflect your thinking. You are about to talk to a lot of people about a project that many of them will think is eccentric. Resistance among your peers will keep them from getting excited about the project. If you reflect optimism about solar energy you will find that it is contagious. Real estate people, bankers, builders, and engineers will be willing to help you more if you feel good about what you're doing. Make sure you make them feel that their advice is important, and you will get important advice. If you plan in a dull manner, you'll probably end up with dull advice and a less than maximized project.

Second, familiarize yourself with your local real estate market. Concentrate your attention on new houses. Know approximate current values of contemporary designs, traditional designs, and any other solar houses. Pick everyone's brains. Talk to appraisers, real estate brokers, etc. While you're out traveling around, look for a competent, adventuresome builder in whom you can put your confidence. Talk to him; be enthused. Builders like to talk about building and the more you know about it the better you will be able to communicate intelligently with him and all the other people you must work with.

The third major step is to talk to several mortgage bankers or savings and loan officers. Qualify yourself financially and discuss appraisals of other solar homes they have financed. Using their advice after they have reviewed your equity position and income, set the upper limit of your budget. Be prepared to spend this much money (or more) because chances are your champagne taste syndrome is about to reach its finest hour. Be honest with yourself. Since there remains a lot of resistance to solar energy homes among bankers, you will probably have to sell your lender. Imagine the additional stature which his bank will derive from financing one of the first solar homes in your area. He might even like to put a sign up on your property. If so, let him. It's good for both of you.

If you are considering building your own home, discuss with your finance people the advantages and disadvantages of such projects versus having a competent builder handle the entire job. Even though builders like to think and act as though they're all getting rich, the 10% to 20% margin of overhead and profit is *earned.* Discuss other owner-built projects the bank has handled and ask for their advice. You never get anything without paying a price and many people underestimate their project's time requirements, and overestimate the amount of money they can save by doing it themselves.

As the fourth step down the road to your solar home, go back to the real estate people and locate a site you *love.* Don't settle for less, but watch your budget. If you overspend here, chances are it will be more difficult for you to build what you want. Help educate your real estate broker on the requirements of a solar energy site. Stay excited about the project and chances are he'll also become enthused on your behalf. Often developers want solar homes in their area as it helps them promote their other homes.

In choosing a site, think sun, sun, sun, south, south, south. Winter wind protection is also desirable. Study the unique area in which you choose to build. Often, microclimates allowing 50% to 75% solar contribution to the heating requirements of the home may be no more than five miles from other zones which could contribute as little as 10%. Several of the books listed in the bibliography provide general climate data; your local weather service is also a good source. Equally worthwhile are discussions with people who already live in similar sun exposures within your area. Talk to others who have built solar homes.

At this point you may need more planning time before making large financial commitments to the project. It is best therefore if you can option a lot for a couple of months, or agree to purchase contingent upon satisfactory engineering studies, satisfactory study of solar energy feasibility, builder's approval, or some other so-called "escape clause." Your ability to buy in this way will depend on how sincere you are with the seller about your intentions. Some sellers will be more patient if they like your idea. The opportunity for a contingent-type purchase also varies with the tightness of the lot market in your area. Your ability to purchase in this manner is, of course, negotiable, but it often allows you the time necessary to research the questions you need answers to as you continue your planning process.

Step five is when this whole thing really starts to become fun. On a sunny day, go out to your site with your spouse or partner. Take with you a book on sun angles, a compass, a couple of chairs, pencil and paper, and a sack lunch. You are in for a full day's project. Study the slope of the lot and sun patterns. Then establish empathy with every rock, tree, blade of grass and structure on and near your site. Meet the neighbors. Challenge your senses to total awareness of everything around you.

Having established rapport with your site, you are ready to pull out pencil and paper and write down *functional requirements* unique to your family. On top of the page write down your upper limit budget. Then write down all the individual spaces your household will require. List the number of bedrooms, utility room, kitchen type and size, formal or informal dining room, family room, den, living room, etc. Then, realizing the need for spatial compromise, write down your prefered *zoning*. Bedrooms in clusters? Kitchen to view the back yard, garage relationship to the kitchen, etc. You now should have established a basic orientation of how you will be living on your site.

Finally, turn on your imagination, tune in your functional requirements, zoning needs, site, family, friends and neighbors and tune out all other problems. Dream freely and fantasize about *how you want your home to feel to you.* High energy sculptural? Laid back funk? Think dimensions of Space/Time, think traditional. Imagine your furniture, your friends over for dinner, think eclectic design, think futuristic, think solar. Move quickly before you lose this inspiration, write it down or better yet talk into a tape recorder. Do this with complete emotional release. Don't quit until you're both tired and you feel mentally strained to express any more feelings. Wow! You did it! You've just decoded everything you need to logically evaluate your design or plan selection parameters. Now go have a nice dinner or go out on the town to celebrate. Chances are you have made decisions which will have a positive effect on the rest of your life. You have on paper a description of a home which should almost completely satisfy not only your basic living needs, but your feelings and emotional needs as well. Don't go on until you've completed this step. *It is the most important step* in home planning if you are to build close to what you *want.* You still have plenty of time to compromise. Now is not the time to be totally practical. Be realistically idealistic.

Try to concentrate this step into a one-day project. Learn to trust your first impressions during this process. They are almost always the best if you are in the right frame of mind. Another good time for hanging loose like this is when you are doing the interior decorating for your home. Stick to your original impulses. If you try to evaluate logically your emotional feelings about your home and lifestyle you will become confused and it will likely scare the hell out of you. Don't look back. Enjoy it and go on from here.

Step six requires that you get back down to earth. Select a custom builder and a solar expert to work with. Pick open-minded people with whom you can discuss the whole project. Evaluate them carefully in terms of their honesty and competence, in that order! If they are the right people for you, they will have your best interests at heart and will show awareness of the fact that they don't have all the answers either. Study other work they've done, both solar and non-solar. If they have happy customers behind them it's likely you'll end up happy, too. Builders and solar engineering people love it when somebody really puts faith in them. This faith can pay big dividends to you during the total building process.

Discuss all your options with your newly acquired experts. Carefully review all the designs in this book. Let your experts help you evaluate all alternatives. Call us at Space/Time if you need clarification (person-to-person for the design consultant please, in Sedona, Arizona (602) 282-3639). After considering all aspects, make a decision, weighing budget, function, and aesthetics. Land on it feet first. Your first impression after weighing all the parameters is probably the best one once again. Place your order for plans, specs, building manual and cost control format using the form in the back of this book. If you need more information on a particular design in our planbook, you may now order one set of complete floor plans excluding structural detailing as an interim planning step.* Just send us a filled out order coupon and a check for 15% of the listed package price. (Ekose'a designs #31, 34 and 35 are excluded from this offer.)

If you find, after going through the design process described here, that you like our designs in general, but that no one plan really suits your needs, Space/Time Designs can be commissioned to execute a custom design specific to your requirements and desires. Our fee for such a service will vary with the complexity of the project. Total cost for a normal project — including design consulting, $1,500.00 retainer, engineering, drafting, travel and other out-of-pocket expenses, and project supervision — should typically run around 6% to 9% of total construction costs.

Step seven starts when you receive your package. Study the plans and specs carefully and work out any remaining

**The cost of this "Interim Planning Set" may then be deducted from the full price if the complete package for that design is subsequently ordered.*

ambiguities. Have your builder, a local building designer, or a structural engineer study the plans and check for local building code requirements. Two blank pages with Space/Time Design title blocks are provided so that you may further customize your final package locally. Earth sheltered designs will include reproduceable structural drawings using our drafting format which your local structural engineer should be able to complete and combine with the building package. Also commission your solar expert to begin to engineer the collector area, storage, and duct work size. Have him consider the design of passive features such as mass quantity, insulating and shading devices, and convective air flows, constantly considering what materials and components are locally available to you.

Our plans cannot possibly be designed for maximum solar efficiency in *all regions of the country.* If you work with someone who has a full understanding of your climate and the thermal principles that determine the quantity of solar aperture, storage, and air flow, refinement of these designs can be accomplished in most areas of the country north of about the 38th parallel (which runs roughly on a line through Washington, D.C., St. Louis, Colorado Springs, and Sacramento). Also note the suggested optimal geographic locations on the thermal envelope designs, plans #31, 34, and 35. Enough details have been included to enable you to roughly evaluate the solar effectiveness of the basic design. After receiving the plans, you may have to adjust collector size, storage size, passive aperture and storage mass, mass wall thickness, air flow passages, duct sizes, insulating and shading devices, etc., to your pocketbook and climate.

Most minor changes can be made by your builder or solar engineer. Redline all sets of drawings so that all of you are clear on what's to be done. By having you and your builder initial all eight sets of plans and specs when you think you've made all necessary decisions, you also will have sets for the subcontractors, carpenters, the building department, and your bank that you know are all alike. If major changes are required for your site or functional needs, we suggest you have them modified by your local building designer or architect.

The building manual for each of our designs includes a control system which, if properly used, triples as a cost estimate, purchasing system, and a cost control format. The manual includes a comprehensive descripion as to how to use it and work with subcontractors and other people involved in the building process. We have used this system with good results for several years. When used realistically and accurately, only minimal deviations have been experienced — usually the result of the unpredictable nature of inflation or labor hours. By subcontracting out the whole project yourself, you potentially can save some money, but *don't* do this unless you know a lot about construction. For certain don't start construction until a very thorough process of cost estimating and subcontractor negotiations has been completed. Other builders and people from other plan services agree that we offer a control system unique in the plan market area. We are now offering it separately from the rest of our material for $45.00. By following our system closely and making sure you have a figure for *every* blank, you or your builder can build your house with a cost deviation from your original cost estimate of from two to five percent. If you do better than this, you've missed your calling; you should be a professional cost estimator.

If you have your builder do the whole job, show him the building manual anyway. It can save money for both of you, increase the quality of the home, and allow for a much more controlled project than you might otherwise have. It can be easily adapted to a cost plus project. At the least it should help you eliminate a lot of guesswork.

Step eight is three-fold. First, commit all contractual arrangements to paper and signatures. If something (either cost or specification item) remains to be decided, identify it and plan to resolve it by a specific time. Don't leave a lot of loose ends. Nobody in building likes changes, so *avoid change orders.* With ambiguities resolved, submit the necessary copies of site plan, specs, and plans to the local building department for structural and code plan checking. At the same time, submit the plans, specs, site plan, contracts, and cost information to your banker and make application for your financing.

If changes are still necessary prior to construction, use the blank pages to provide additional detailing. From the time you get our package until you start construction, you should budget between $200 and $700 for local engineering (both solar and structural) and designer drafting time. Earth sheltered designs in this book will likely run higher and the availability of a structural engineer knowledgeable about concrete shell design and theory should be considered before ordering Spacedome I or Spacedome II building packages. This will be a function of how many changes you make from our drawings, but often good advice and accurate detailing early in the game will save you heavy expenses later.

Pay close attention to detailing the contract documents. Most problems in building a home are due to misunderstandings. Nearly all "builder horror stories" are the result of poor builder/client communication leading to poor subcontractor/employee/contractor/client communication, leading to at least four sets of irrational, defensive, and fearful people worrying about themselves instead of about building a house which *was* the job they all originally set out to do, all with the best of intentions.

The last and final step after you've closed on your site, secured financing, and made all contractual arrangements is to take a short rest. You'll need it, as building a house is often something like an endurance contest.

Bruce Anderson, in the preface of his book, *Solar Energy — Fundamentals In Building Design* states, "Somehow I just can't help thinking that solar energy is going to make it this time. Twenty years ago public interest in solar peaked due to rising fuel prices and predicted energy shortages, then waned due to solar's relatively high cost and to intense Federal interest in nuclear. Energy prices are low relative to the cost of solar, and tenacious Federal support for nuclear continues unabated; but this round, fortunately, the American public knows better. We know that energy prices are being held down artificially and that they could skyrocket when political forces see it to be in their best interest. We know enough not to be so easily seduced by the nuclear hard sell. We are at last realizing that the world's pantry of non-renewable natural resources does indeed have a bottom shelf; that we had better be damned frugal with what little remains; and that we had better turn to less conventional sources to provide for our reduced needs."[3]

Since the first edition of *Let's Reach for the Sun* was published in 1978, much has transpired within our American consciousness with respect to energy consumption and alternative resources. The Middle East political wars and unrest have made the possibility of oil cutoffs imminent. Inflation, high interest rates, and dire predictions for more of the same eat away at the American dream of economic security and home ownership. The great irony of this economic scenario is that the construction industry is the one worst hurt in tough economic conditions, while, with current innovations in building design and thermal technology, it has the greatest potential of all to help American homeowners become self-sufficient in energy *and* lifestyle. How is it possible that a supposedly evolved society has put fiscal and monetary fantasy ahead of the earth's natural balance and our children's future? Can we continue the policies of immediate self gratification which cause soured water and air and visual blight, not to mention nuclear proliferation, and still survive as a free and self-sufficient society? These questions must be answered *soon.*

Perhaps it is time for all of us to take a long hard look at history. It has been said that the Roman Empire fell because of moral decay. It has also been theorized that they succumbed to economic deterioration because they decimated the forests of northern Italy for industrial fuel. It seems possible that these two theories may be mutually inclusive; what could be more immoral than to destroy the only free blessing we share . . . the natural beauty and resources of the earth. Civilization today could be facing a similar fate . . . only this time, if we could only see, we have the technical knowhow to avoid such a fall. Simply by building in the most self-sufficient way possible you can take a step toward responsible self government and help us all avoid the potential tyranny and decay inherent in the primary alternative, nuclear power.

Before you begin studying our plans in the following section, I have some good news and some bad news. The bad news first. The solar industry does not have *all* the answers yet. Now the good news. If you have read this material, studied the suggested reading list, and are keeping current on solar periodicals, you are probably more knowledgeable about solar space heating than anyone else you know. You should be ready to approach intelligent decisions about your own solar building project. Best of luck from the Space/Time Gang.

Let's reach for the sun!

3. Bruce Anderson, *Solar Energy Fundamentals of Building Design*, New York: McGraw-Hill Book Company, 1977, p. vii.

BIBLIOGRAPHY

AIA Research Corporation. *A Survey of Passive Solar Buildings*. February, 1978.

Bruce Anderson. *Solar Energy — Fundamentals in Building Design*. New York: McGraw-Hill Book Company, 1977.

Bruce Anderson. *The Solar Home Book — Heating, Cooling and Designing with the Sun*. Harrisville, New Hampshire: Cheshire Books, 1976.

Robert Bennet. *Sun Angles for Design*. Bala Cynwyd, Pennsylvania: Robert Bennett, 1978.

Quinton M. Bradley/James F. Carlson. *Solar Primer One — Solar Energy in Architecture — A Guide for the Designer*. Whittier, California: Solarc, 1975.

Stu Campbell. *The Underground House Book*. Charlotte, Vermont: Garden Way Publishing Co., 1980.

Richard L. Crowther. *Sun Earth*. Denver, Colorado: Crowther/Solar Group, 1977.

John A. Duffie/William A. Beckman. *Solar Energy Thermal Processes*. New York: John Wiley & Sons, 1974.

Energy Research and Development Administration. *National Program for Solar Heating and Cooling of Buildings, Project Data Summaries, Vol. 1 — Commercial and Residential Demonstrations*. Washington, D.C., 1976.

Rick Fisher/Bill Yanda. *The Food and Heat Producing Solar Greenhouse — Design, Construction, Operation*. Santa Fe, New Mexico: John Muir Publications, 1976.

Sigfried Giedion. *Space, Time and Architecture*. Cambridge, Massachusetts: Harvard University Press, 1967.

John Hayes/Drew Gillet, editors. *Proceedings of the Conference on Energy-conserving, Solar-heated Greenhouses*. Held at Marlboro College, Marlboro, Vermont: Marlboro College, 1977.

Joint Economic Committee Congress of the United States. *The Economics of Solar Home Heating*. Washington, D.C.: U.S. Government Printing Office, 1977.

John Keyes. *Harnessing the Sun to Heat Your House*. Dobbs Ferry, New York: Morgan & Morgan, 1974.

Jim Leckie/Gil Masters/Harry Whitehouse/Lily Young. *Other Homes and Garbage — Designs for Self-sufficient Living*. San Francisco, California: Sierra Club Books, 1975.

Edward Mazria. *The Passive Solar Energy Book*. Emmaus, Pennsylvania: Rodale Press, 1979.

James C. McCullogh. *The Solar Greenhouse Book*. Emmaus, Pennsylvania: Rodale Press, 1978.

National Solar Heating and Cooling Information Center. *Passive Design Ideas for the Energy Conscious Builder*. Rockville, Maryland.

Alador Olgyay/Victor Olgyay. *Solar Control and Shading Devices*. Princeton, New Jersey: Princeton University Press, 1957.

Roaring Fork Resource Center, edited by Gregory E. Franta/Kenneth R. Olson. *Solar Architecture*. Ann Arbor, Michigan: Ann Arbor Science, 1978.

Robert L. Roy. *Underground Houses: How to Build a Low Cost Home*. New York: Sterling Publishing Co., 1979.

Norma Skurka/Jon Naar. *Design for a Limited Planet — Living with Natural Energy*. New York: Ballantine Books, 1976.

Solar Energy Industries Association. *Solar Industry Annex*. Washington, D.C., 1977.

Solaron Corporation Solar Energy Systems. Denver, Colorado: Solaron Corporation, 1977.

S. V. Szokolay. *Solar Energy and Building*. New York: Halsted Press, 1975.

U.S. Department of Commerce, National Weather Records Center. *Selective Guide to Climatic Data Sources*. Asheville, North Carolina, 1969.

U.S. Department of Housing and Urban Development. *Intermediate Minimum Property Standards for Solar Heating and Domestic Hot Water Systems, Volume 5*. Washington, D.C., 1977.

U.S. Department of Housing and Urban Development, Office of Policy Development and Research. *Solar Heating and Cooling Demonstration Program — A Descriptive Summary of HUD Cycle 3 Solar Residential Projects*. Washington, D.C., Summer, 1977.

U.S. Department of Housing and Urban Development, prepared by Regional and Urban Planning Implementation, Inc. *Home Mortgage Lending and Solar Energy*. Cambridge, Massachusetts, 1977.

University of Minnesota. *Earth Sheltered Housing Design*. Minneapolis, Minnesota: The University of Minnesota, 1978.

Alex Wade/Neal Ewenstein. *30 Energy-Efficient Houses . . . You Can Build*. Emmaus, Pennsylvania: Rodale Press, 1977.

Louis Wampler. *Underground Homes*. Gretna, Louisiana: Pelican Publishing Co., 1978.

Donald Watson. *Designing and Building a Solar House — Your Place in the Sun*. Charlotte, Vermont: Garden Way Publishing Co., 1977.

Malcolm Wells. *Underground Designs*. Brewster, Massachusetts, 1977.

Malcolm Wells and Irwin Spetgang. *How to Buy Solar Heating without Getting Burned*. Emmaus, Pennsylvania: Rodale Press, 1978.

David Wright. *Natural Solar Architecture — A Passive Primer*. New York: Van Nostrand Reinhold Company, 1978.

Periodicals

Solar Age, published monthly by SolarVision, Inc., Church Hill, Harrisville, New Hampshire 03450.

Solar Energy Intelligence Report, published weekly by Business Publishers, Inc., P.O. Box 1067, Silver Spring, Maryland 20910.

Solar Engineering Magazine, published monthly by Solar Engineering Publishers, Inc., 8435 Stemmons Freeway, Suite 880, Dallas, Texas 75247.

Underground Space, published bi-monthly by Pergamon Press Inc., Fairview Park, Elmsford, New York 10523.

Space Time Designs inc. PRESENTS

30 ORIGINAL SOLAR AND EARTH SHELTERED HOME DESIGNS

By ordering a "Complete Package" for one of the following designs, you will receive:

- One 18'' x 24'' architectural rendering of the home
- 8 sets of ¼'' scale drawings which will include:
  - All floor plans (2 or 3 pages)
  - Foundation plan (1 page)
  - Floor framing plan (1 page)
  - Roof framing plan (1 page)
  - All four elevations
  - Necessary cross sections
  - Details, fireplace faces, cabinet elevations, railing details and other miscellaneous details
  - Electrical plans
  - 2 blank pages of tracing paper with Space/Time Designs, Inc. title blocks for additional local customized detailing
- One building manual which includes advice on subcontractor selection, cost estimating and cost control during construction
- One detailed estimating/cost control format which includes some quantity surveys
- One set of suggested building specifications plus one set of blank specs
- One copy of a suggested builder/owner contract

By ordering the "Interim Planning Package"* at 15% of the complete package price, you will receive:

- One 18" x 24" architectural rendering of the home
- One set of ¼" scale drawings which include all the above pages except electrical plans

*Note: Cost of the "Interim Planning Package" may subsequently be deducted from the full price if the complete package for that design is ordered later.

THE OVEN HOUSE

S/T-1

With almost a four-to-one ratio of floor area to collector area, this building has proven to be an effective energy conserver. As built in Seattle and slightly improved for this book, this home boasts a very simple, livable floor plan, and an artistically pure design. Vertical collectors reflect and multiply features from the surrounding landscape, presenting a kaleidoscope of images to the passerby.

From a ground level entry, a winding stairway leads to living areas and the luxurious master suite. A round mirror that opens to reveal a window is detailed over the dressing vanity. A conveniently located second bedroom could double as a den or sewing room if desired. In the living room, a geometric fireplace design echoes the surprising wood-trimmed circular opening between living and dining areas. Across the hall from the dining room an efficient kitchen opens to the breakfast nook. Downstairs a private guest bedroom claims its own bath and the game room shares in the warm light from the sunken greenhouse.

Flat-mounted water pre-heat collectors over the entry, underground siting of the north side of the lower level, a steep north roof slope, an energy-efficient fireplace, and super insulation make this home a working example of solar effectiveness.

Main floor — 1,515 sq. ft.
Lower floor — 900 sq. ft.
Greenhouse volume — 1,050 cu. ft.
Total square footage — 2,415

Package price — $510.00
3 bedrooms
3 bathrooms

Passive Features
Greenhouse glazing — 84 sq. ft.
Potential mass storage — 150 cu. ft.

Active Features
Potential collector area — 600 sq. ft.
Potential storage volume — 400 cu. ft.
Collector tilt — 90°
Water pre-heat system

180

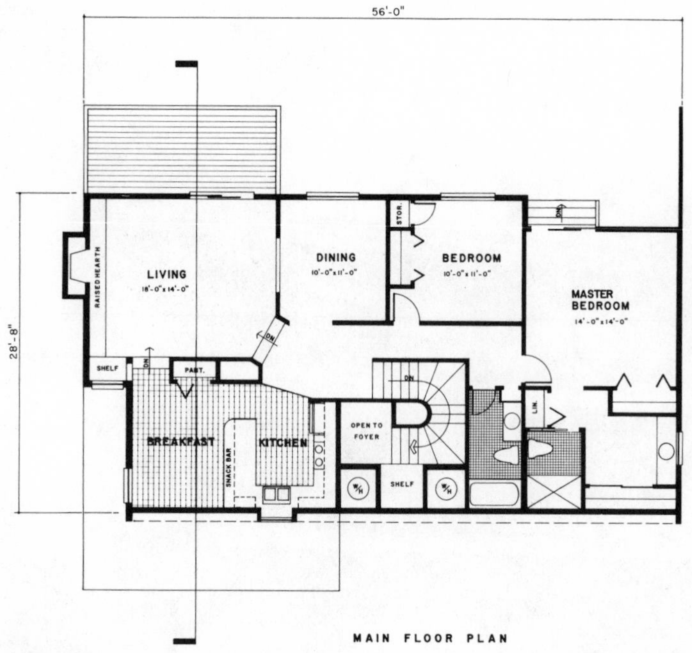

MAIN FLOOR PLAN

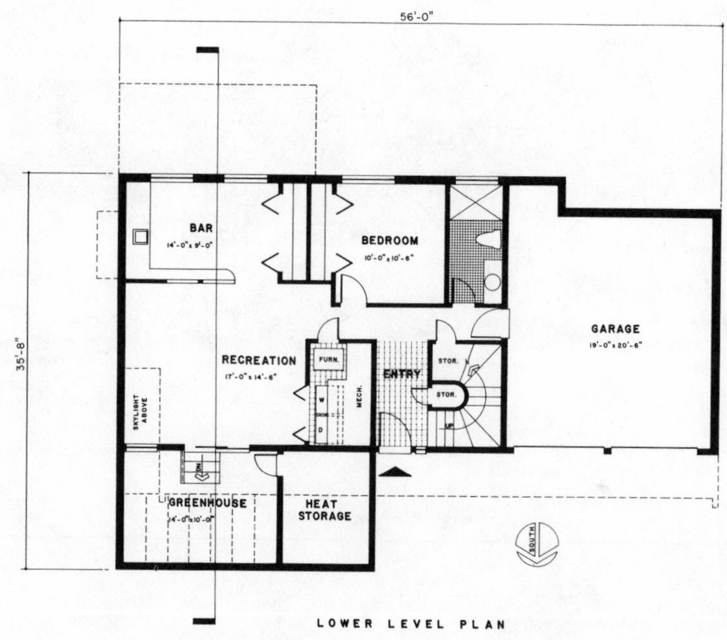

LOWER LEVEL PLAN

Space Time Designs inc.

S/T 1 - THE OVEN HOUSE

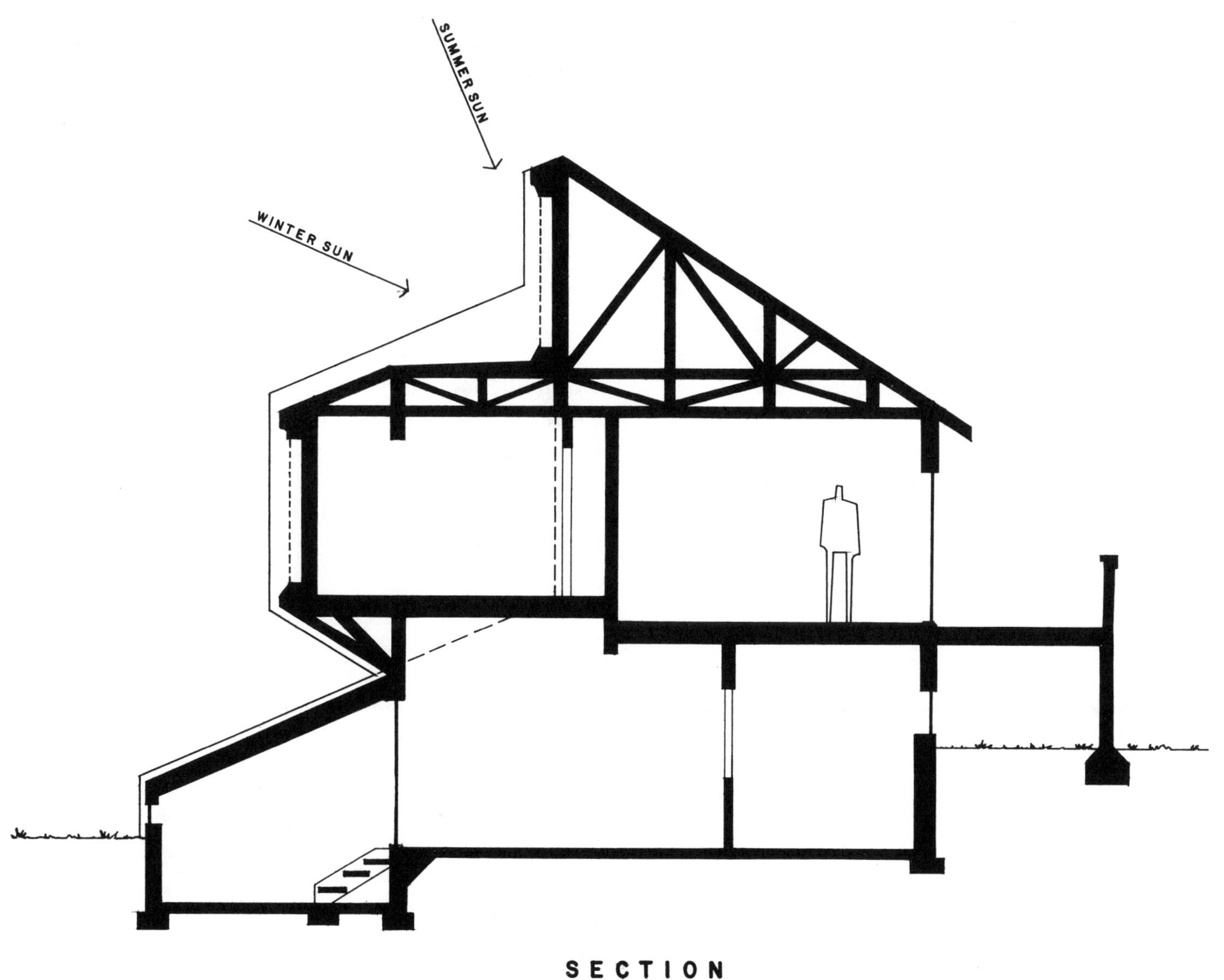

SECTION

WEST OF THE MOON
S/T-2

Conceived in a spirit of fairytale fantasy, this delightful multi-dimensional design by Dave Fordon will make an enviable showcase for any plant freak's collection. The simple and convenient floor plan features a living room surrounded by a virtual botanical garden and private studio half a level below the living area which will be alight with morning sun. Enjoy summer dining on the cool north-facing deck or serve meals buffet style to a gang from the eating bar between kitchen and cozy family room.

Upstairs, you can stay warm all night without the furnace by opening bedrooms to the stored warmth of the garden lounge or by stoking the centrally-located circulating fireplace before going to bed. Skylit bathrooms and an openable dormer ensure both light and comfort. Although not shown here, a detached garage is included with the plans.

The exterior features pyramidal windows, skylights with adjustable louvres, a custom detailed front door and rustic wood siding. A reflective surface south of the vertical-mount roof-top collectors will give this home an effective winter collector area in excess of 275 square feet.

Main floor — 1,470 sq. ft.
Upper floor — 887 sq. ft.
Total square footage — 2,357

Package price — $610.00
2 bedrooms plus study
2½ bathrooms

Passive Features
Direct gain glazing — 234 sq. ft.
Potential mass storage — 400 cu. ft.

Active Features
Potential collector area — 192 sq. ft.
Potential storage volume — 150 cu. ft.
Collector tilt — 90°
Water pre-heat system

SOLTNER

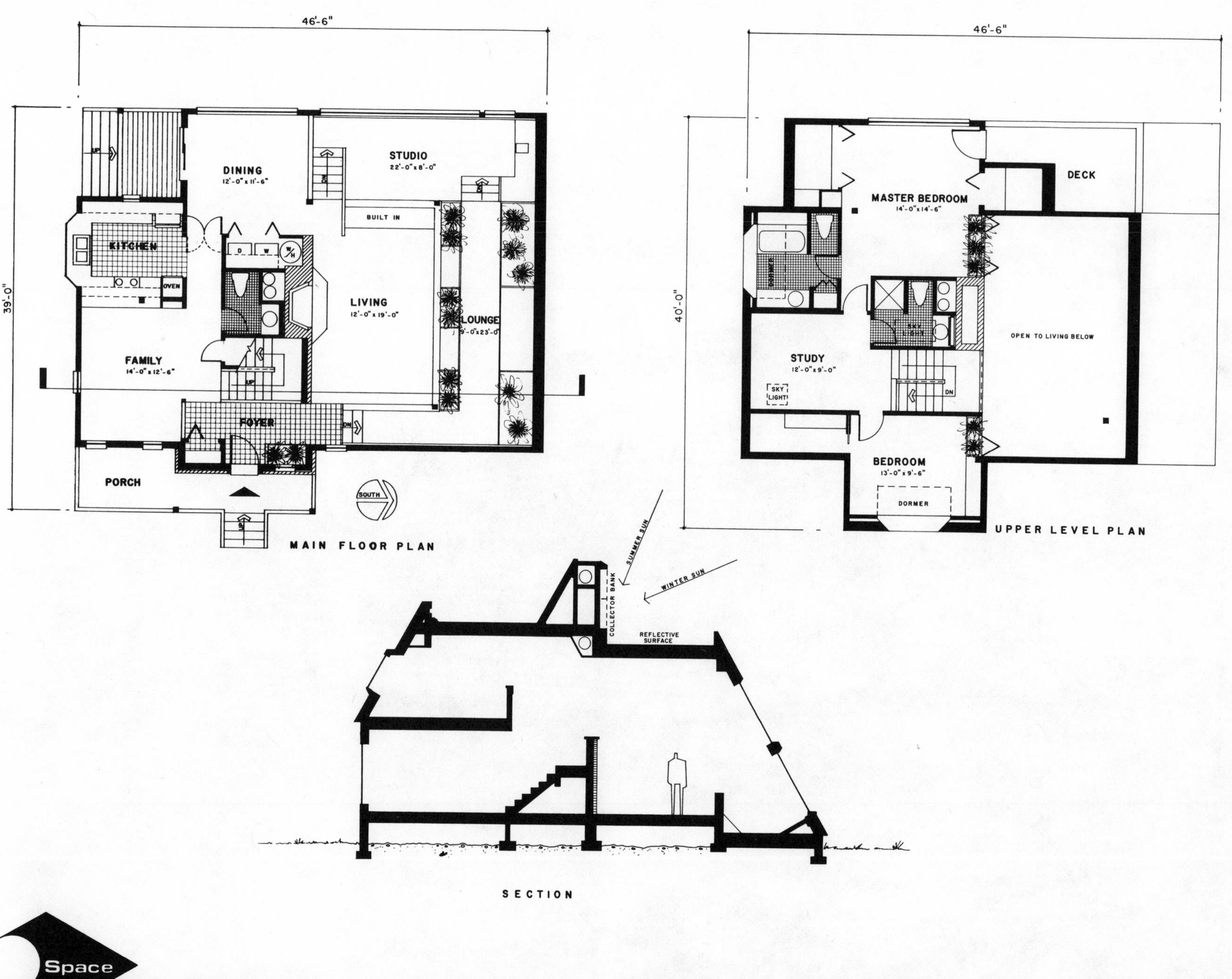

Space Time Designs inc.

S/T 2 - WEST OF THE MOON

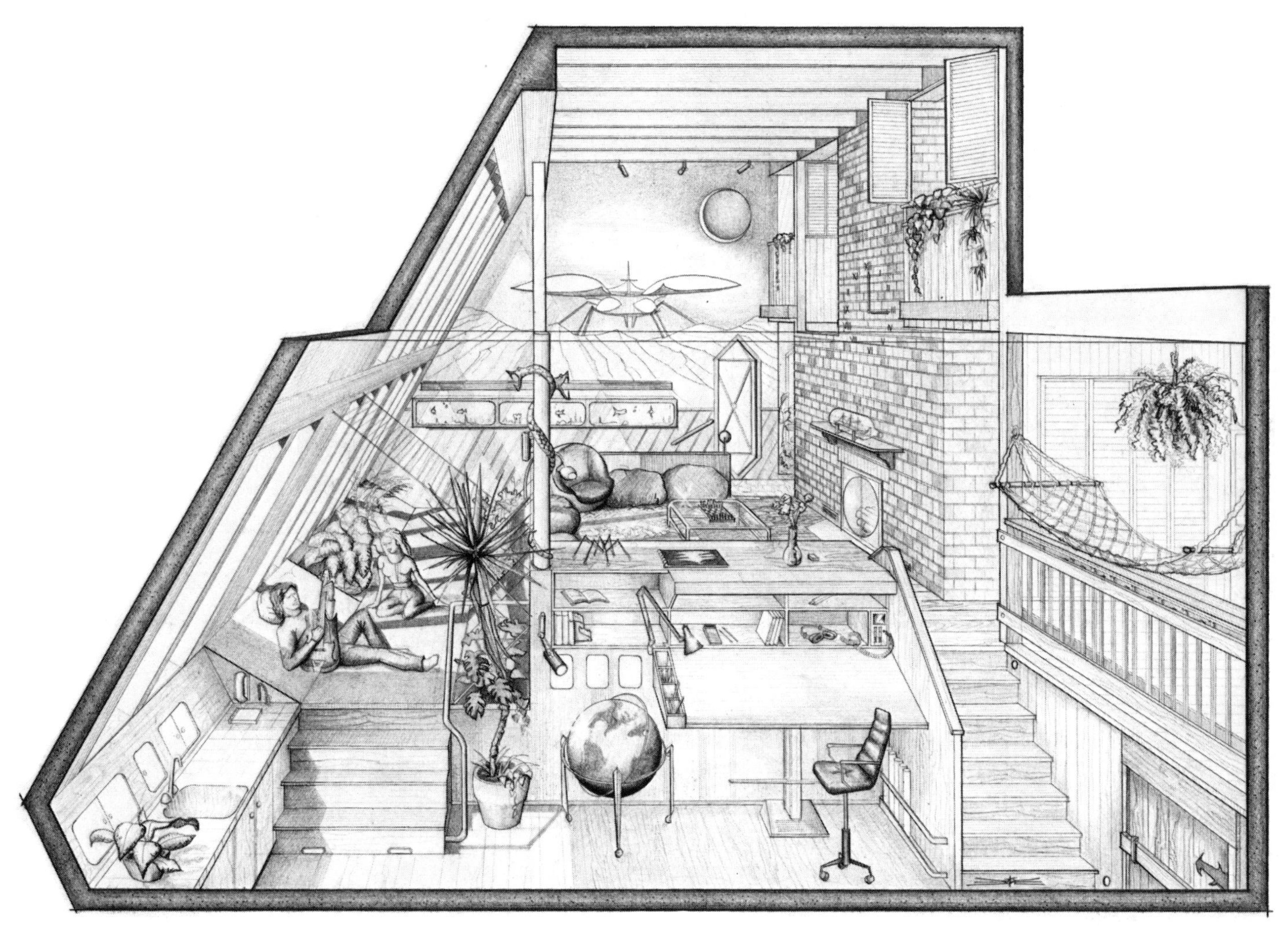

SUN BUCKET
S/T-3

Simplicity of construction methods and adaptability to existing surroundings are two of the primary advantages of this eyepleasing design. Entering through an airlock lit by the swirling colors of a round stained glass window, you ascend to a centrally located informal dining area which visually extends beyond the sliding glass door into a huge two-story greenhouse. A combined sitting and living area designed for informal entertaining may be visually separated by strategic placement of furnishings, and the flight of stairs at the south end leads you down into the lower level of the greenhouse. The compact kitchen with a generous snack bar features its own greenhouse deck for plants that require close attention. Across the hallway you find the master bedroom with its large walk-in closet and continental bath. The lower level features two more bedrooms with centrally located bath and utility area.

Now step outside the greenhouse and imagine yourself viewing the entire south elevation with its two-story greenhouse, plants thriving within, balconies and stairs, the active collector bank above, and the soaring shape 30 feet high reaching for the sun, all within an economical 1,800 square foot split-level.

Main floor — 1,256 sq. ft.
Lower floor — 600 sq. ft.
Greenhouse volume — 5,544 cu. ft.
Total square footage — 1,856

Package price — $280.00
3 bedrooms
2 bathrooms

Passive Features
Greenhouse glazing — 555 sq. ft.
Potential mass storage — 1,000 cu. ft.

Active Features
Potential collector area — 273 sq. ft.
Potential storage volume — 336 cu. ft.
Collector tilt — 60°
Water pre-heat system

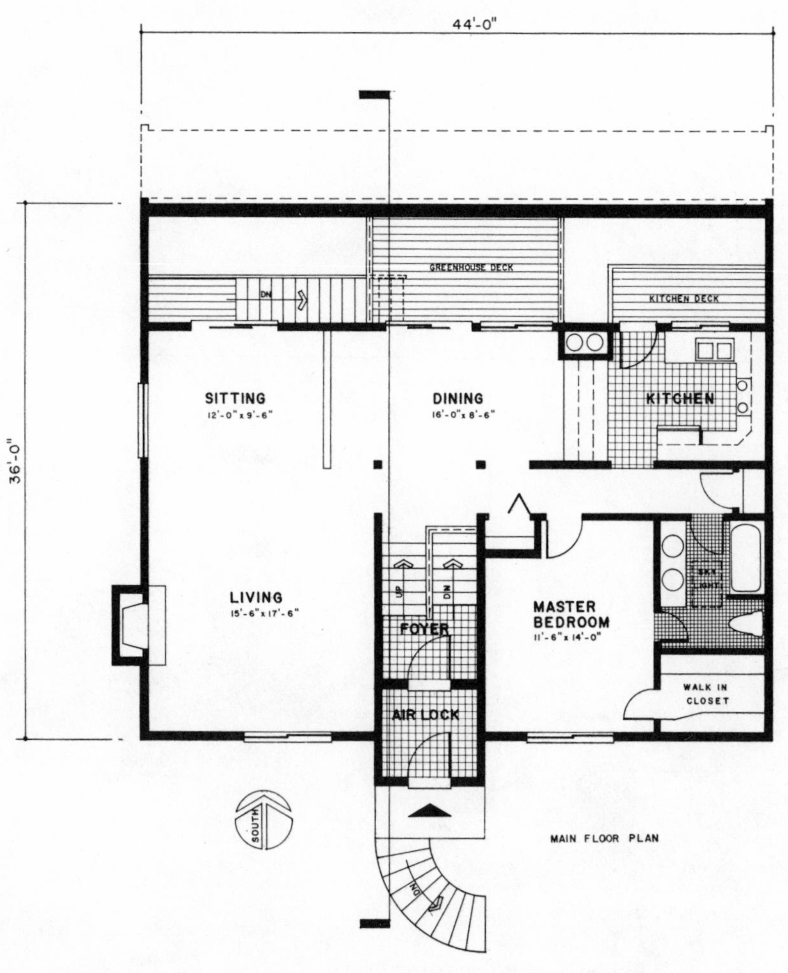

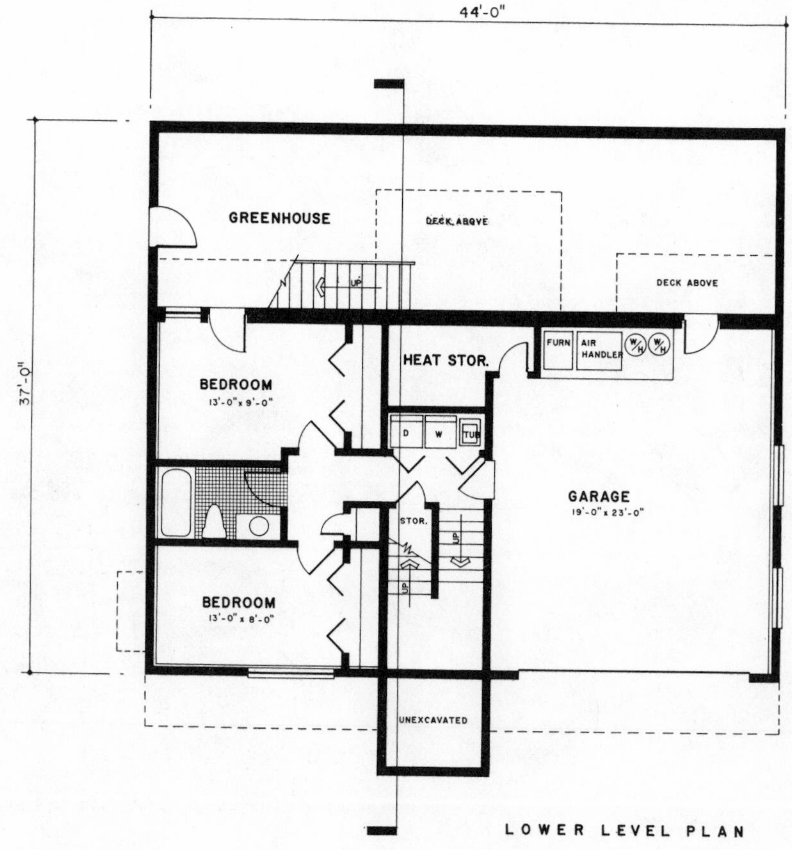

Space Time Designs inc.

S/T 3 - SUN BUCKET

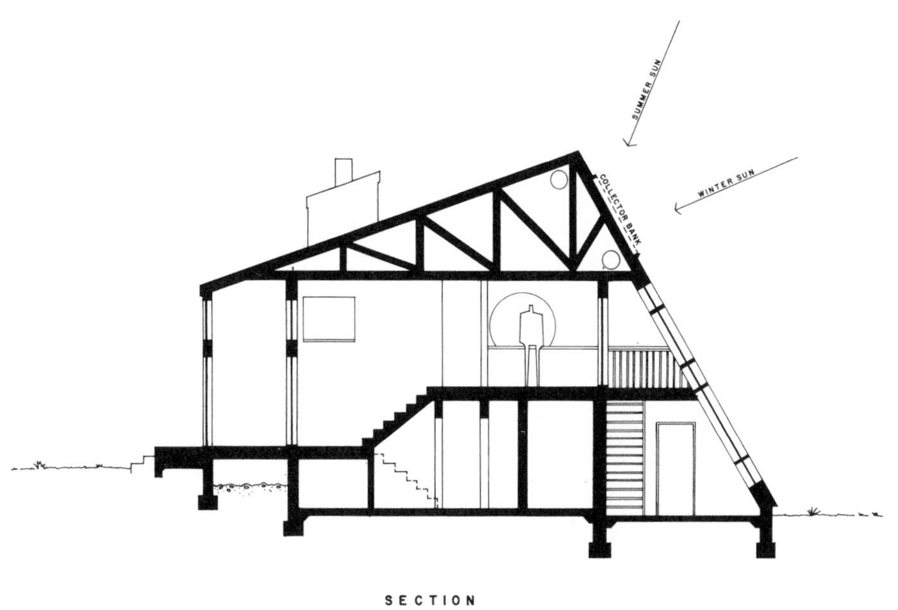

SECTION

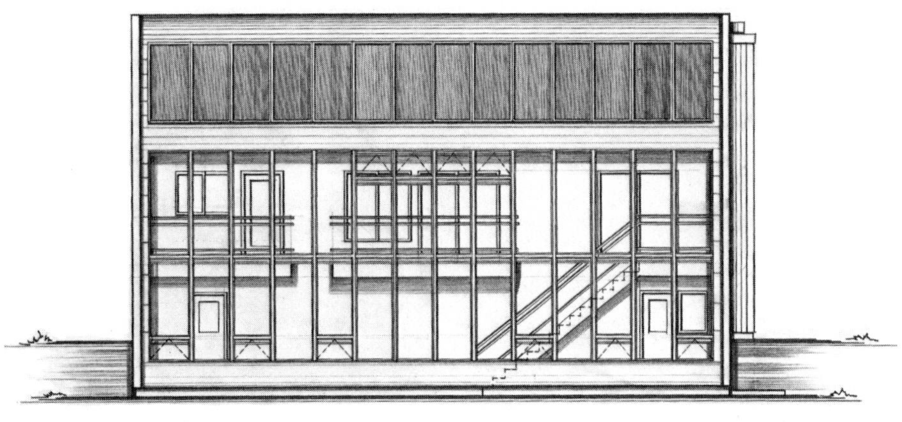

SOUTH ELEVATION

SOLAR HOUSE ON THE PRAIRIE
S/T-4

Have you ever wanted your own ranch out on that spread that gives you room to roam and the cozy atmosphere of rustic country living surrounded by exposed beams and rough-sawn woods? Yes? Then "howdy partner," let me show you around.

Bring the gang up the covered porch, and come into the wide entry hall where a massive open staircase ascends squarely around a huge wooden post. Through its open treads you discover the dining area with a high rough-sawn wood ceiling and dark-tiled, direct gain floor slab. The island kitchen, wide open to the family area, provides room for country cooking and a bright, sun-warmed space for relaxed living with family and friends. Through a sliding door and a large window you can watch your favorite plants thrive in the attached greenhouse, and the second greenhouse on the other side of the sundeck will provide a view of the back forty and an appropriate backdrop for the heat-efficient, airtight Franklin stove facing the beamed living room.

The two large upstairs bedrooms will be warmed at night and in cloudy weather by the heat collected and stored in the water barrels above the greenhouses. The nearly windowless north wall and open design of the plan will allow extremely efficient use of the solar energy collected by this down home ranch house.

Main floor — 1,246 sq. ft.
Upper floor — 756 sq. ft.
Greenhouse volume — 1,568 cu. ft.
Total square footage — 2,002

Package price — $685.00
3 bedrooms
2½ bathrooms

Passive Features
Direct gain glazing — 90 sq. ft.
Greenhouse glazing — 243 sq. ft.
Potential mass storage — 500 cu. ft.

Active Features
Water pre-heat system

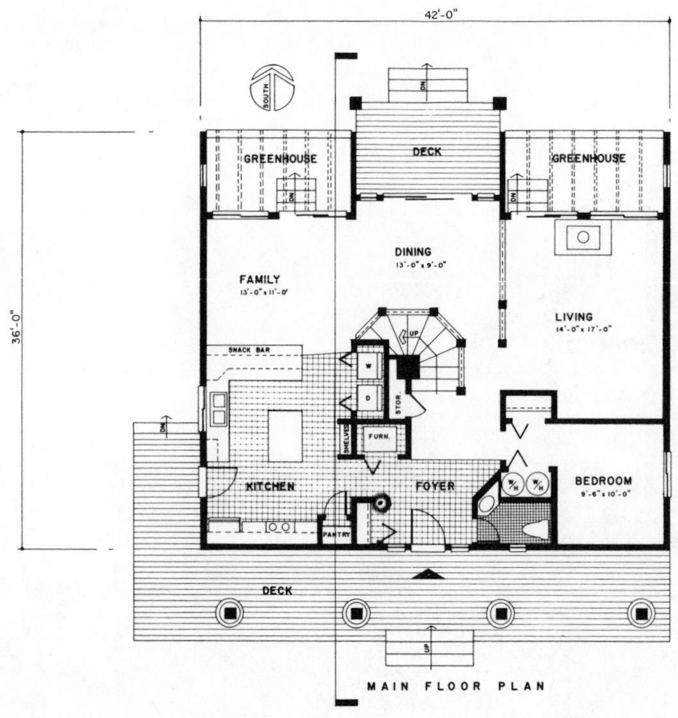

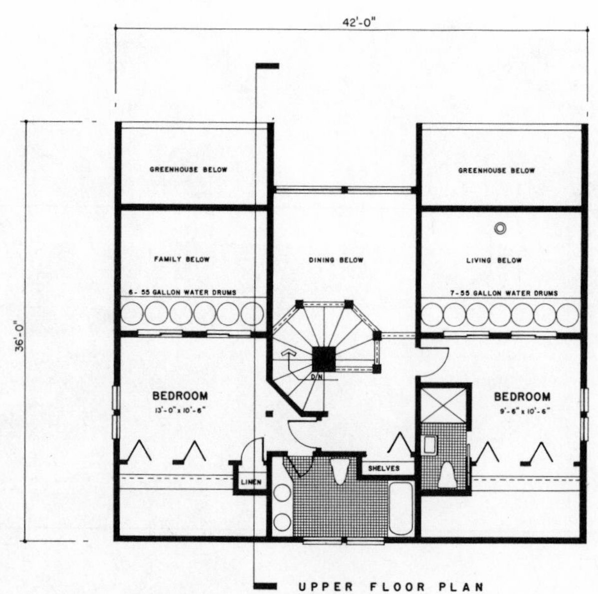

S/T 4 - SOLAR HOUSE ON THE PRAIRIE

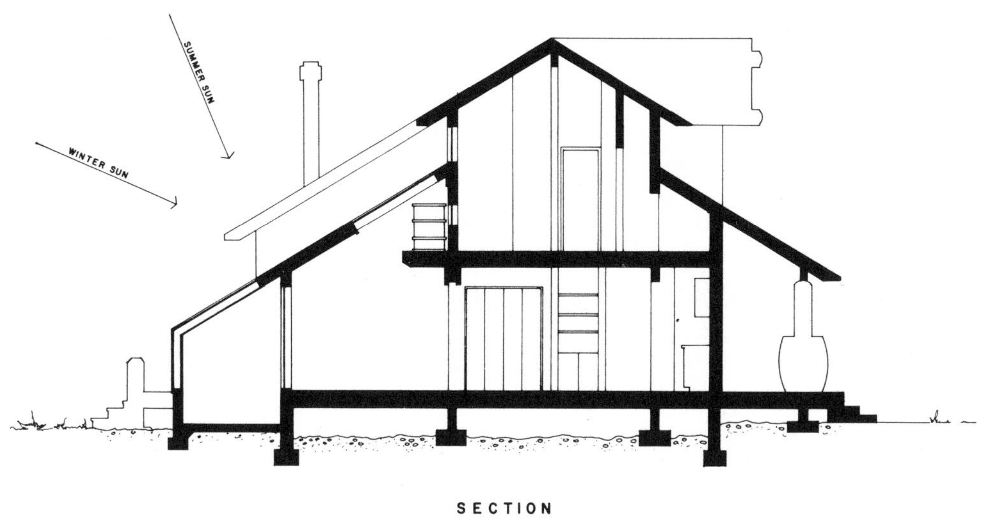

SECTION

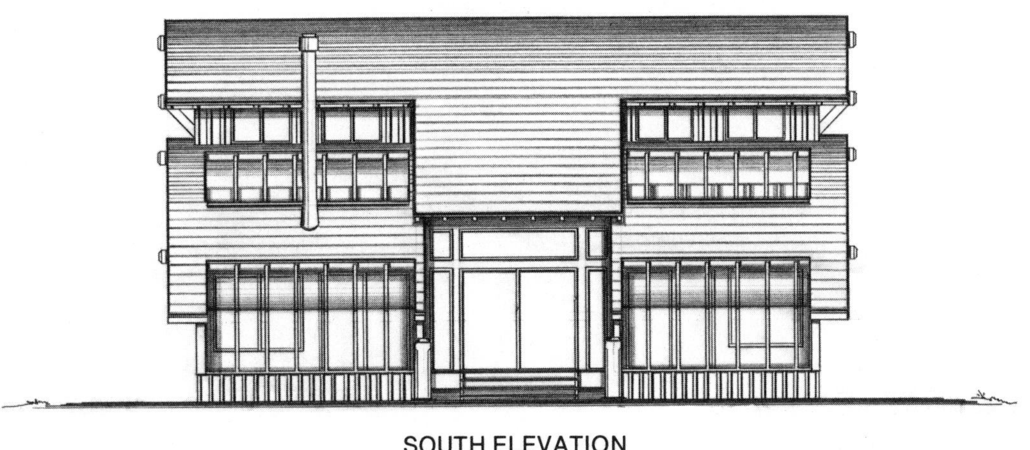
SOUTH ELEVATION

JACKSONVILLE
S/T-31

The uniquely designed spaces of this light and airy home will afford the ultimate in gracious southern living. Connected by the warm and lofty spaces of the cedar-trimmed central galleria, the two wings neatly separate the living and sleeping spaces. Living and dining areas and kitchen all partake of the light and spaciousness of the main galleria, while off the back of the living area is a cozy, low-ceiling fireplace nook, perfect for relaxing with a good book and your favorite music. The large kitchen features one whole wall of handsome wood shelving for displaying a prized basket or pottery collection. The huge utility room is close to both kitchen and rear entry for maximum efficiency and convenience.

And while part of the family is entertaining in the living module, others can be sleeping or studying in the sleeping area. The large master bedroom, which shares the light and airiness of the secondary greenhouse with the second bedroom, features a large walk-in closet and a luxurious tile shower in the master bath. The two secondary bedrooms share a second large bath. Solar hot water pre-heat panels operate efficiently in the conditioned space of the secondary solarium.

As in other thermal envelope designs for humid climates, east and west cooling walls and an underground preconditioning system are required. Designed to be reminiscent of the open and light-filled spaces of a Japanese home, Jacksonville should be an ideal house for year-around comfort during both warm and cool Southern weather.

House square footage — 1,915
Greenhouse volume — 8,857 cu. ft.

Package Price — $550.00

3 bedrooms plus study
2½ bathrooms

Passive Features

Exterior vertical greenhouse glazing — 690 sq. ft.
Greenhouse 45° roof glazing — 93 sq. ft.
Geo-thermal air envelope system

Active Features

Optional water pre-heat system

CASTILLO '80

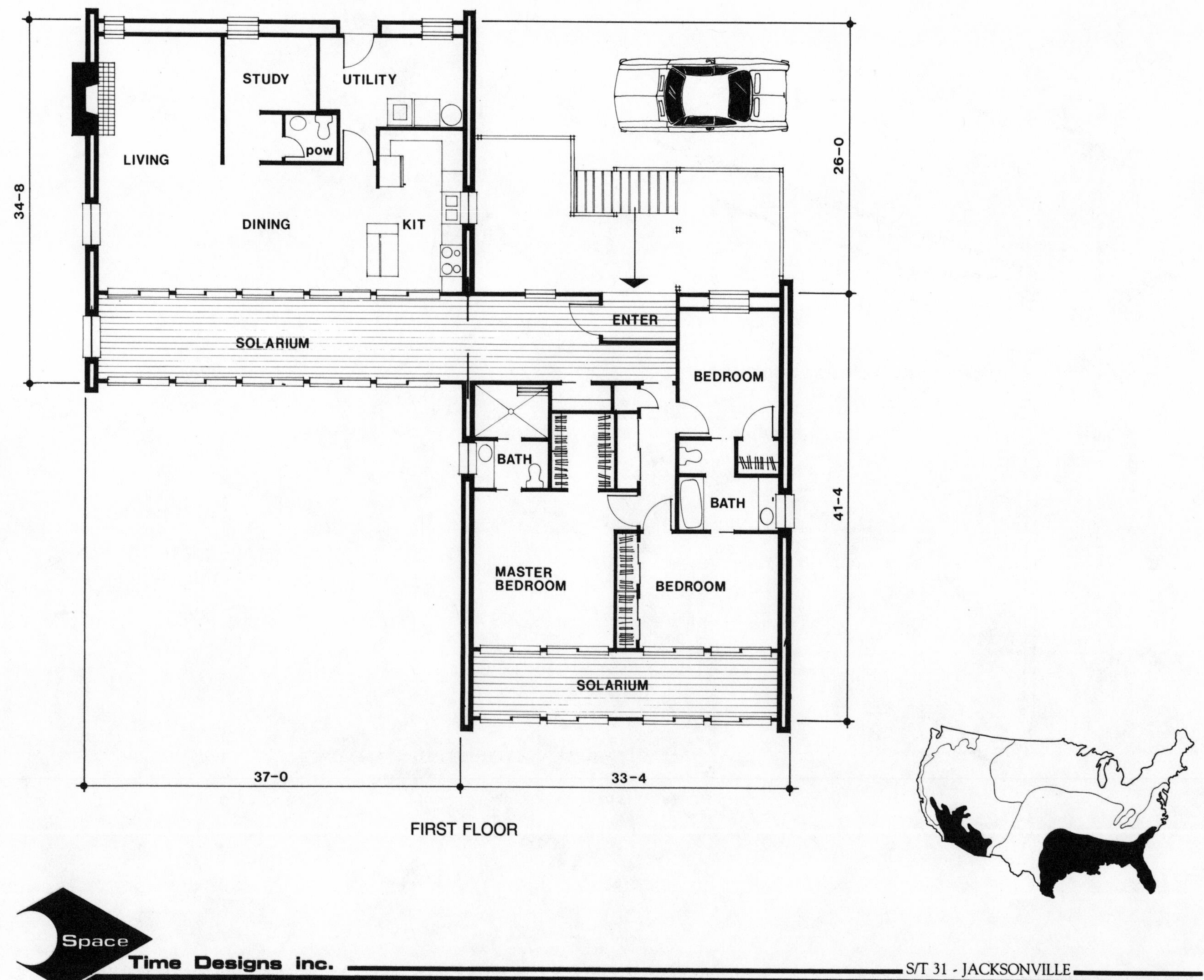

FIRST FLOOR

Space Time Designs inc.

S/T 31 - JACKSONVILLE

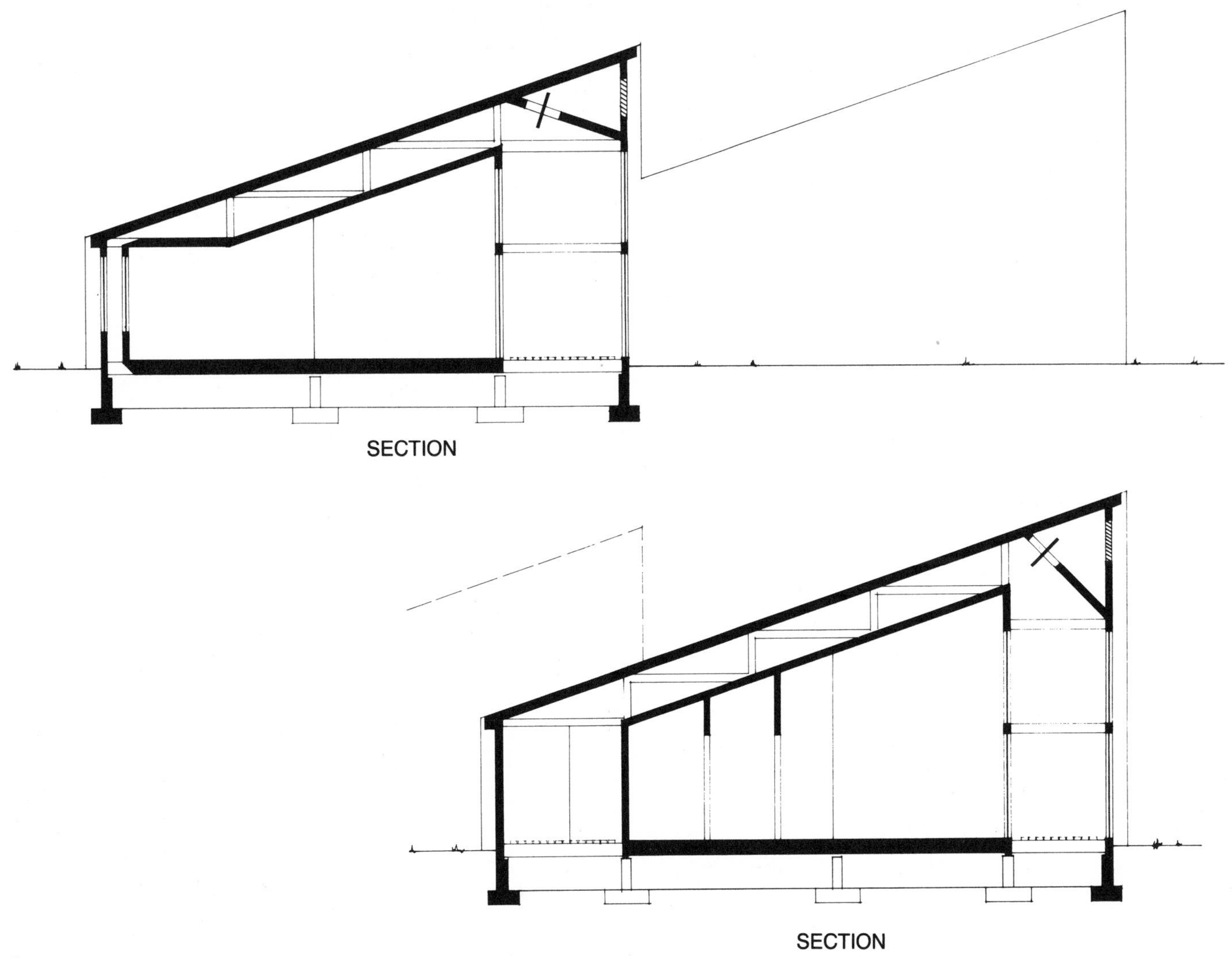

SECTION

SECTION

WINGS
S/T-6

This uni-space vacation home designed by Dave Ellis combines the ambiance of wood with the sculptural qualities of solar design. Warmth and comfort are provided by a two-story mass wall that will establish a double convection loop. The freestanding fireplace in the sunken, sky-domed conversation pit may very well be the only back-up heating unit necessary in many areas. The overall amount of solar assistance desired can be adjusted simply by increasing or decreasing the size of the glazed mass walls.

A large peninsula kitchen is visually separated from the dining area by the ascending spiral of the staircase which leads to the sleeping area — a large bunk room and the master suite with its private bath. The two sleeping areas open to the conversation pit below and benefit from the warmth of both fireplace and mass wall.

Detailed in post and beam construction, the home promises carefree and relaxed weekends. The pure geometrics of the wedge wall provide an ideal deflector for north winter winds while the towering glass facade on the south will admit ample light. In addition to the heat radiated by the mass wall in winter, cooling in summer is effected by taking advantage of the solar chimney effect — drawing cool north air through the living spaces to the vents at the base of the mass wall and exhausting air out the top vents.

Main floor — 895 sq. ft.
Upper floor — 704 sq. ft.
Total square footage — 1,599

Package price — $375.00
2 bedrooms
2 bathrooms

Passive Features
Direct gain glazing — 126 sq. ft.
Mass wall glazing — 504 sq. ft.
Potential mass storage — 424 cu. ft.

Active Features
Water pre-heat system

SOLTNER

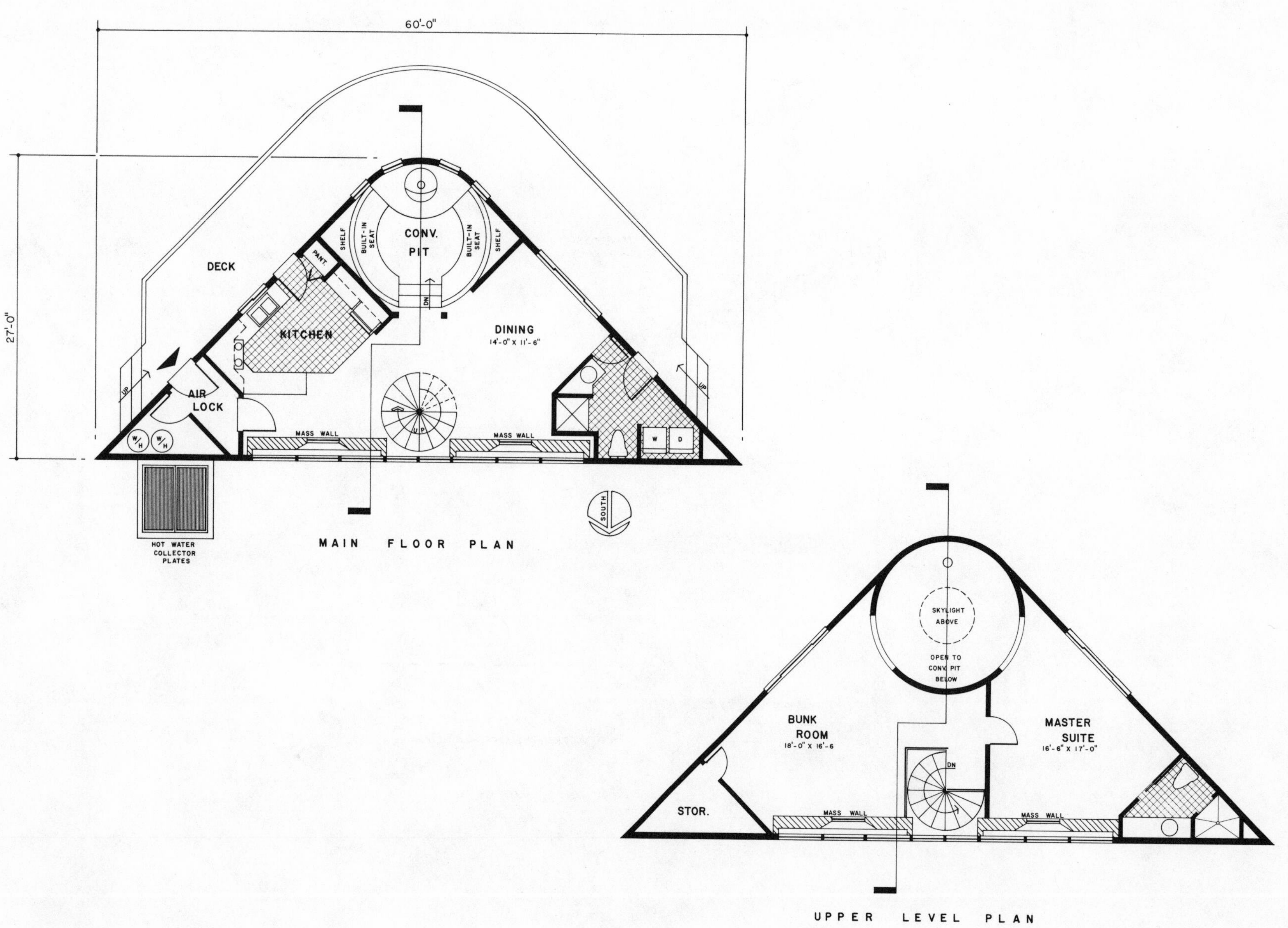

Space Time Designs inc.

S/T 6 - WINGS

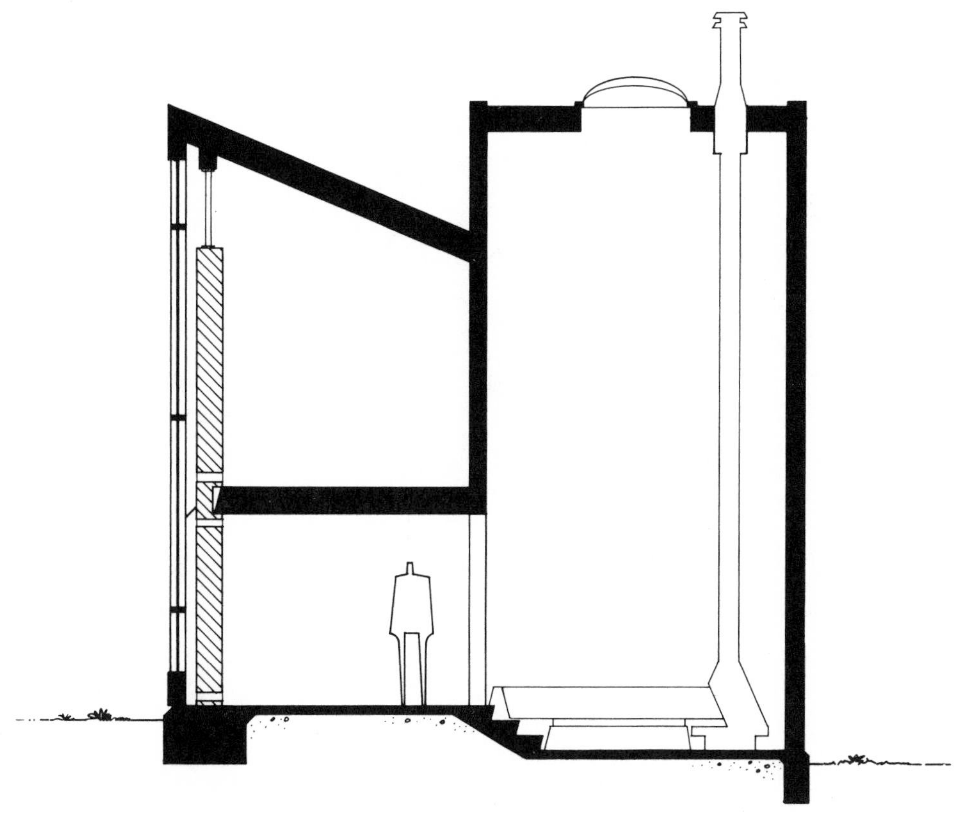

SECTION

RADIO SOLAR MUSIC HALL
S/T-7

Whether built as a view beach house, a forest tree house, or a city dwelling, this home emanates openness and tranquility. Enter at the dormitory level (containing two bedrooms, large bath, and utilities) and ascend the stairway lit by glowing panes of leaded glass to the open complexity of the main living area. Picture a grand piano in the music area, a quiet gathering around the freestanding fireplace, a dinner next to the soaring, multi-level greenhouse.

Upstairs, the master bedroom opens in surprising ways to the spaces below and offers a secluded studio with north light for painting or study, as well as its own penthouse access to the greenhouse.

Vertical collectors make this home especially efficient if snow followed by clearing is a frequent weather pattern in your area. Reflected radiation can add 15% to 30% solar effectiveness over the optimum tilt planes in the late winter months in northern latitudes. A large south patio or reflective driveway could also add solar input. Hydronic collectors atop the south-facing roof, located off the master bedroom, is one suggested water pre-heat system. Garage design as shown on S/T-16: *Old Green Thumbs* is included in plans but not shown.

Main floor — 794 sq. ft.
Lower floor — 824 sq. ft.
Upper floor — 596 sq. ft.
Greenhouse volume — 1,881 cu. ft.
Total square footage — 2,214

Package price — $460.00
3 bedrooms plus studio
2 bathrooms
Detached garage included but not shown

Passive Features
Greenhouse glazing — 216 sq. ft.
Potential mass storage — 450 sq. ft.

Active Features
Potential collector area — 518 sq. ft.
Potential storage volume — 384 cu. ft.
Collector tilt — 90°
Water pre-heat system

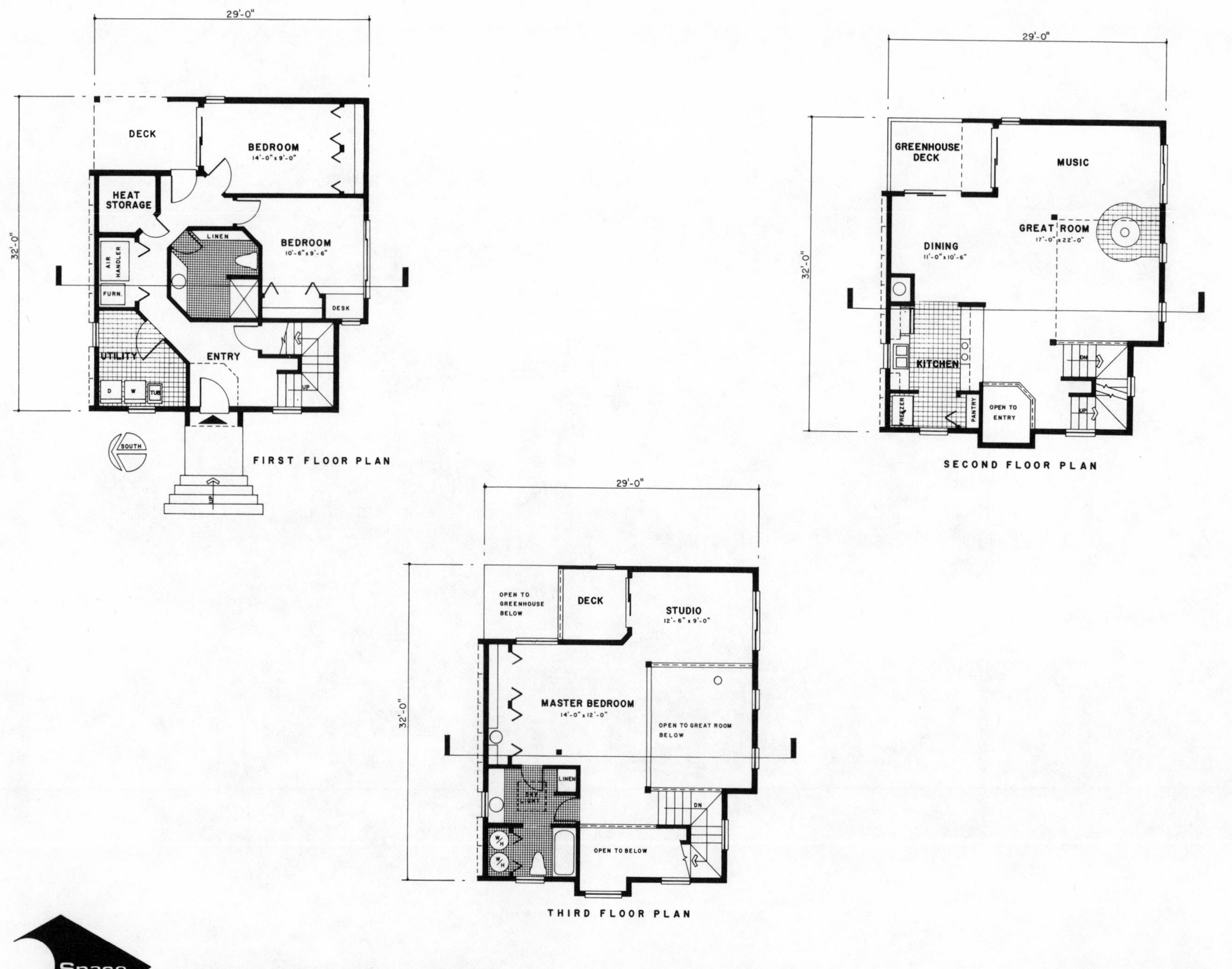

Space Time Designs inc.

S/T 7 - RADIO SOLAR MUSIC HALL

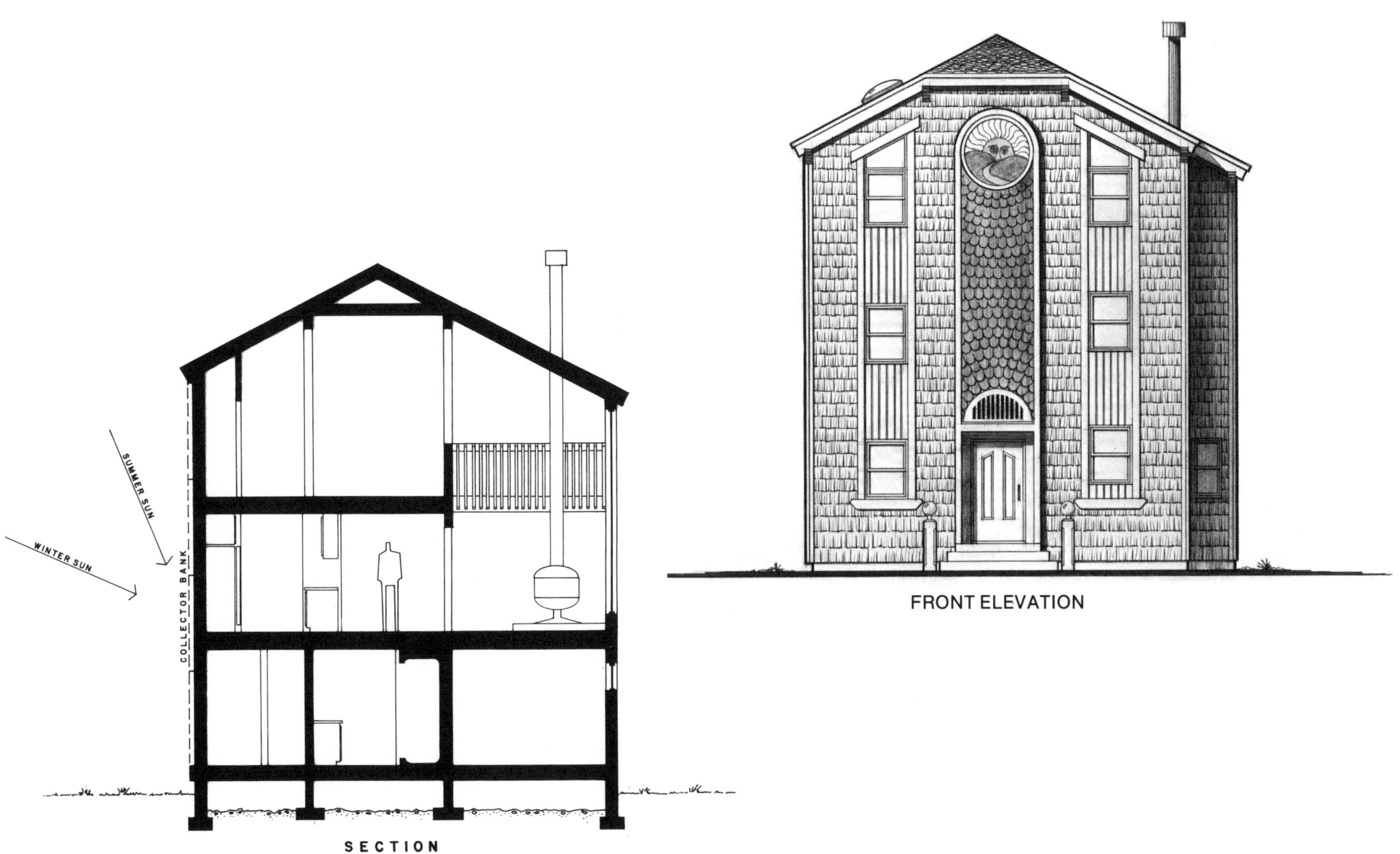

SECTION

FRONT ELEVATION

SUN BATH
S/T-8

This compact, spatially surprising house combines open living spaces with maximum solar effect. A steep north roof slope and the curving stair tower deflect winter winds while the nearly windowless front invites you into a skylit airlock entry, and then into the exciting living room dominated by the curve of the seating area around the fireplace. The glazed mass wall behind the fireplace serves as heat storage for both greenhouse and fireplace. The angled south collector wall creates dramatic interior spaces in both dining room and greenhouse. A second mini-greenhouse enhances the spacious kitchen, centrally located to be convenient to the garage and utilities, as well as to the main living spaces.

Up the skylit circular staircase, its shape echoing the fireplace seating area, there are two compact bedrooms with special niches for desks and dressers, and the spacious master suite.

The sculptural shape and flowing curves of Sun Bath should fit well into most contemporary neighborhoods.

Main floor — 1,089 sq. ft.
Upper floor — 1,075 sq. ft.
Greenhouse volume — 1,680 cu. ft.
Total square footage — 2,164

Package price — $340.00
3 bedrooms
2½ bathrooms

Passive Features
Greenhouse glazing — 252 sq. ft.
Inner mass wall glazing — 120 sq. ft.
Potential mass storage — 350 cu. ft.

Active Features
Potential collector area — 317 sq. ft.
Potential storage volume — 308 cu. ft.
Collector tilt — 60°
Water pre-heat system

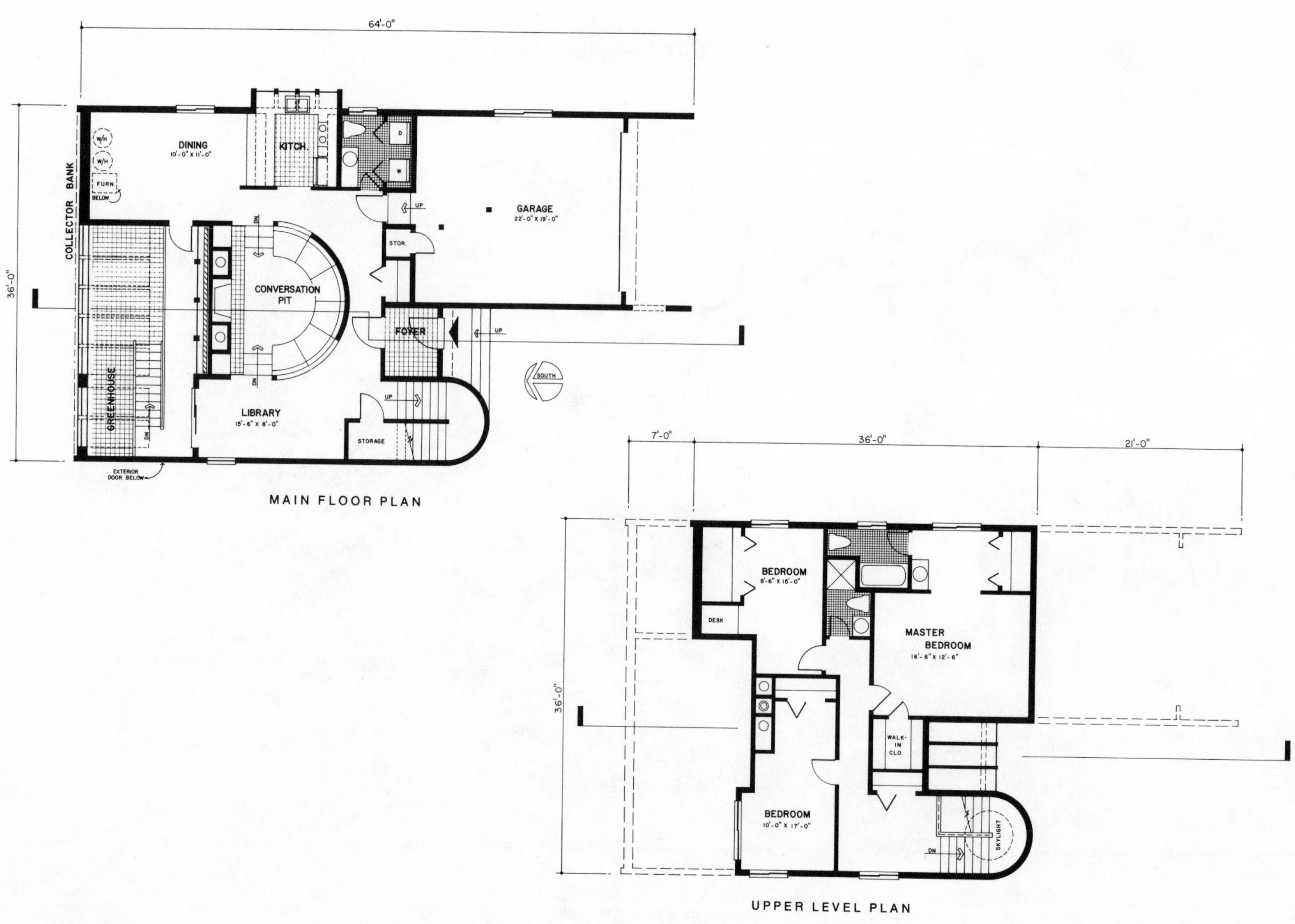

Space Time Designs inc.

S/T 8 - SUN BATH

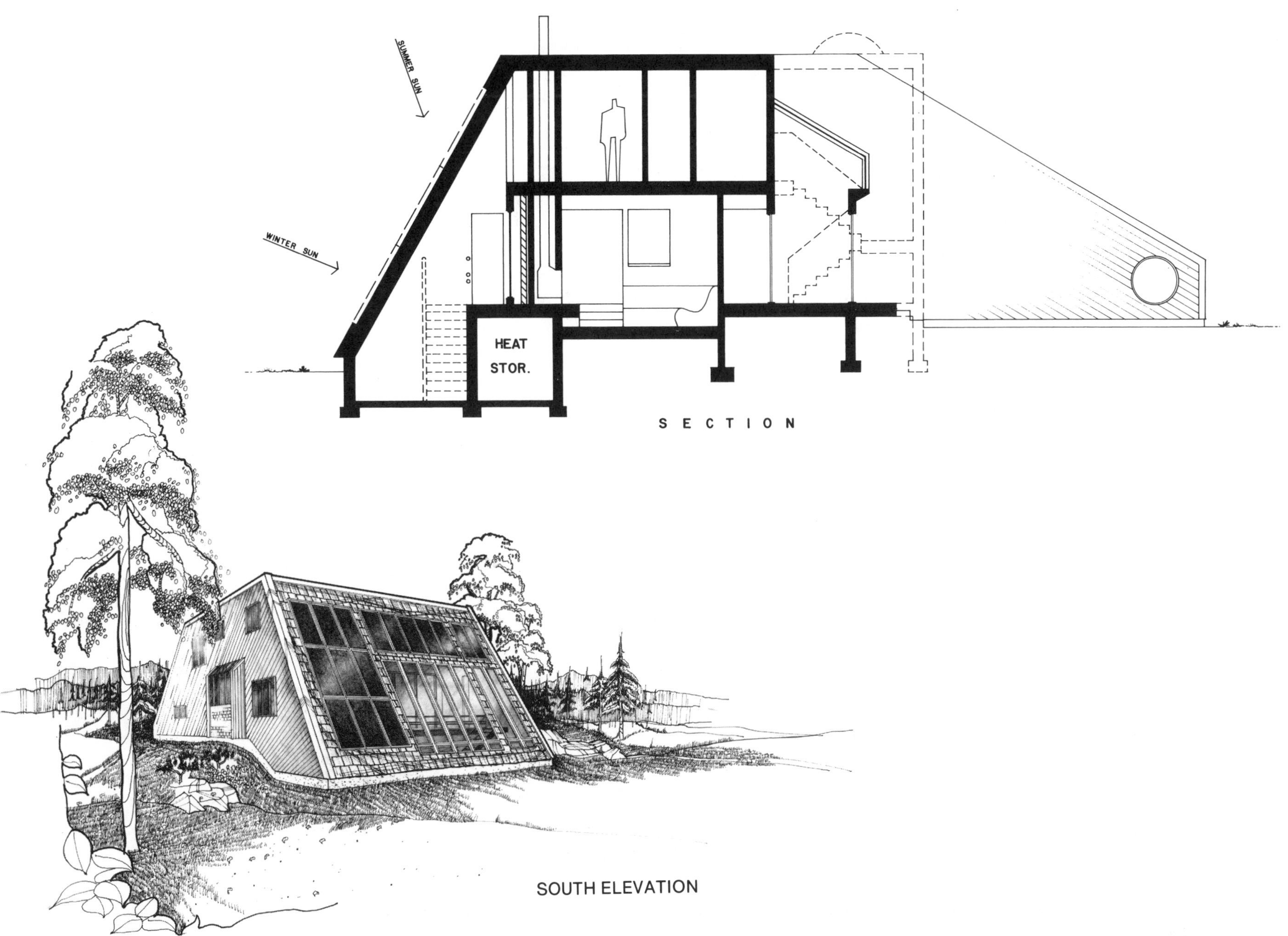

SOUTH ELEVATION

EARTHWORM
S/T-32

Live in an earth sheltered, solar assisted home that will blend into your neighborhood so well that "not even your banker will know for sure." Designed with a nice balance between the thermal protection of earth cover and optimum winter solar gain, this house will be surprisingly light, warm and spatially gratifying . . . and comparatively easy to build too.

Protected from fierce winter weather by nearly 60% earth cover, the design is ideal for any steep, south-facing slope.

Imagine your dinner guests' excitement as they enter Earthworm through a warm, plant-filled solar entry greenhouse, accept your hospitality before an open fire before being entertained in the circular sky-domed dining area. Sunlight warms the bedroom areas and warm air rises to the clerestory to be circulated by a high furnace return or stored in a massive concrete interior wall for later circulation. The high clerestories are openable for summer cooling and allow the sun to illuminate the hall and master suite.

The combination of exciting contemporary appeal, energy penny pinching, and construction simplicity should make Earthworm an outstanding investment for custom builders and homeowners alike. Structural engineering for earth sheltered areas of the design must be designed and supervised by a local licensed engineer. For this purpose, the plan package will include reproduceable structural pages for his needs. Above-ground garage plans included. Live like an earthworm, and feel free as a bird!

House square footage — 1,483
Garage square footage — 440

Package Price — $540.00
3 bedrooms plus sitting area
2 bathrooms

Passive Features
Direct gain glazing — 210 sq. ft.
Potential direct mass storage — 150+ cu. ft.
Structural mass storage — 1,368 cu. ft.
60% earth sheltered

Active Features
Water pre-heat system
High recirculating warm air return

PETERSON 80

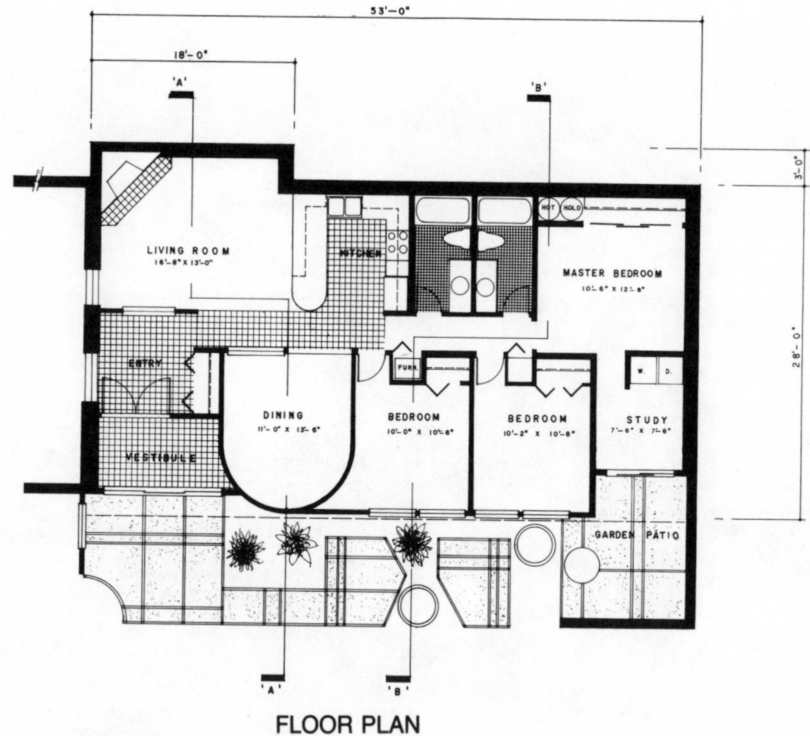

FLOOR PLAN

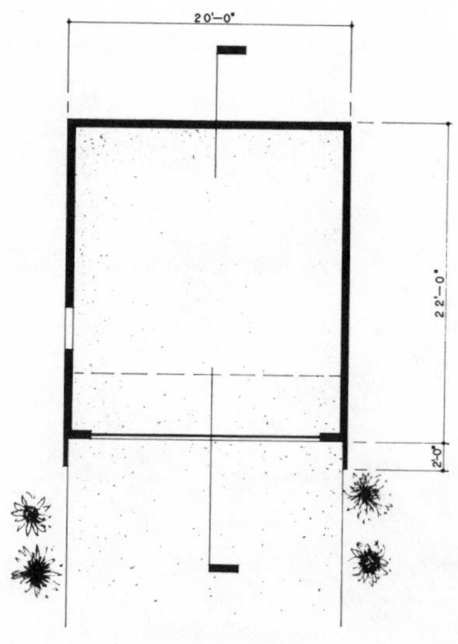

GARAGE PLAN

Space Time Designs inc.

S/T32 - EARTHWORM

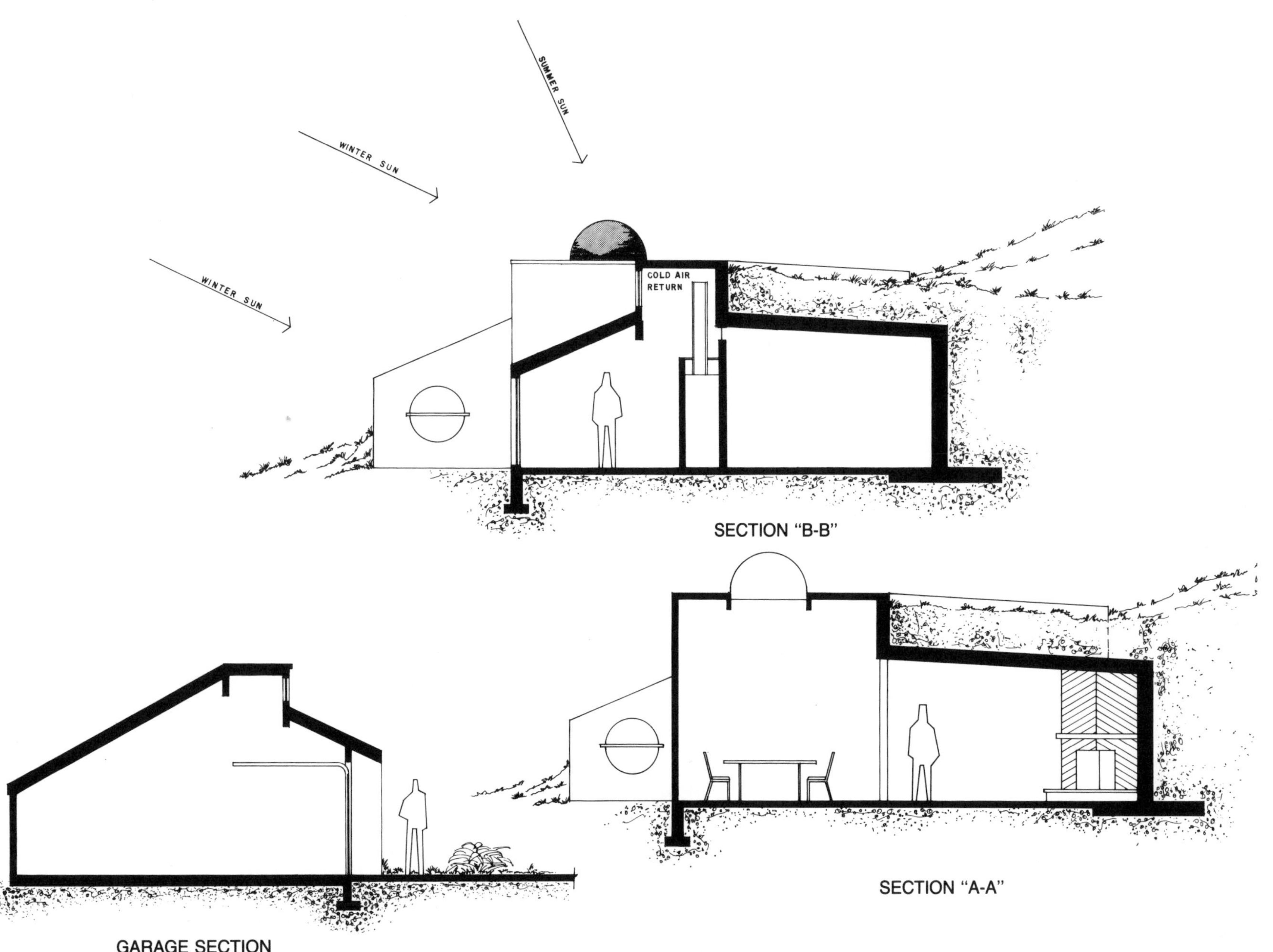
SUMMER SUN
WINTER SUN
WINTER SUN
COLD AIR
RETURN
SECTION "B-B"
GARAGE SECTION
SECTION "A-A"

GREENHOUSE GALLERY
S/T-10

A handsome house from any angle, this design by Keith Soltner can be adapted to a variety of site conditions just by changing the location of the garage and driveway approach.

The glazed mass wall at the back of the greenhouse will circulate heat to the living areas by establishing a convection loop which will supplement the convective loop existing in the greenhouse. With ample room for the addition of mass storage material, this greenhouse will be able to hold stable temperatures over long durations, and if insulated shading devices are used, it will serve as an effective heat sink as well as a visual enhancement to the living space which wraps around it. In addition to heat from the greenhouse and mass wall, both living and family rooms share the actual and psychological warmth of their fireplaces with the living spaces above. A half bath and laundry area are convenient to the spacious, centrally-located island kitchen.

Up the sky-domed radial stairwell, two interestingly-shaped bedrooms adjoin a large playroom. The master bedroom overlooks both the greenhouse and the living room fireplace.

Summer cooling and ventilation is effected by opening clerestory windows and the vents on the outside top of the mass wall to pull air from north windows through the house. With careful insulation and planning for ample heat storage, this home with its massive greenhouse will be ideal for an energy-conscious gardener.

Main floor — 1,166 sq. ft.
Upper floor — 1,025 sq. ft.
Greenhouse volume — 3,744 cu. ft.
Total square footage — 2,191

Package price — $780.00
3 bedrooms
2½ bathrooms

Passive Features
Direct gain glazing — 110 sq. ft.
Greenhouse glazing — 570 sq. ft.
Mass wall glazing — 510 sq. ft.
Potential mass wall storage — 450 cu. ft.
Potential greenhouse mass — 1,300 cu. ft.

Active Features
Water pre-heat system

SOLTNER

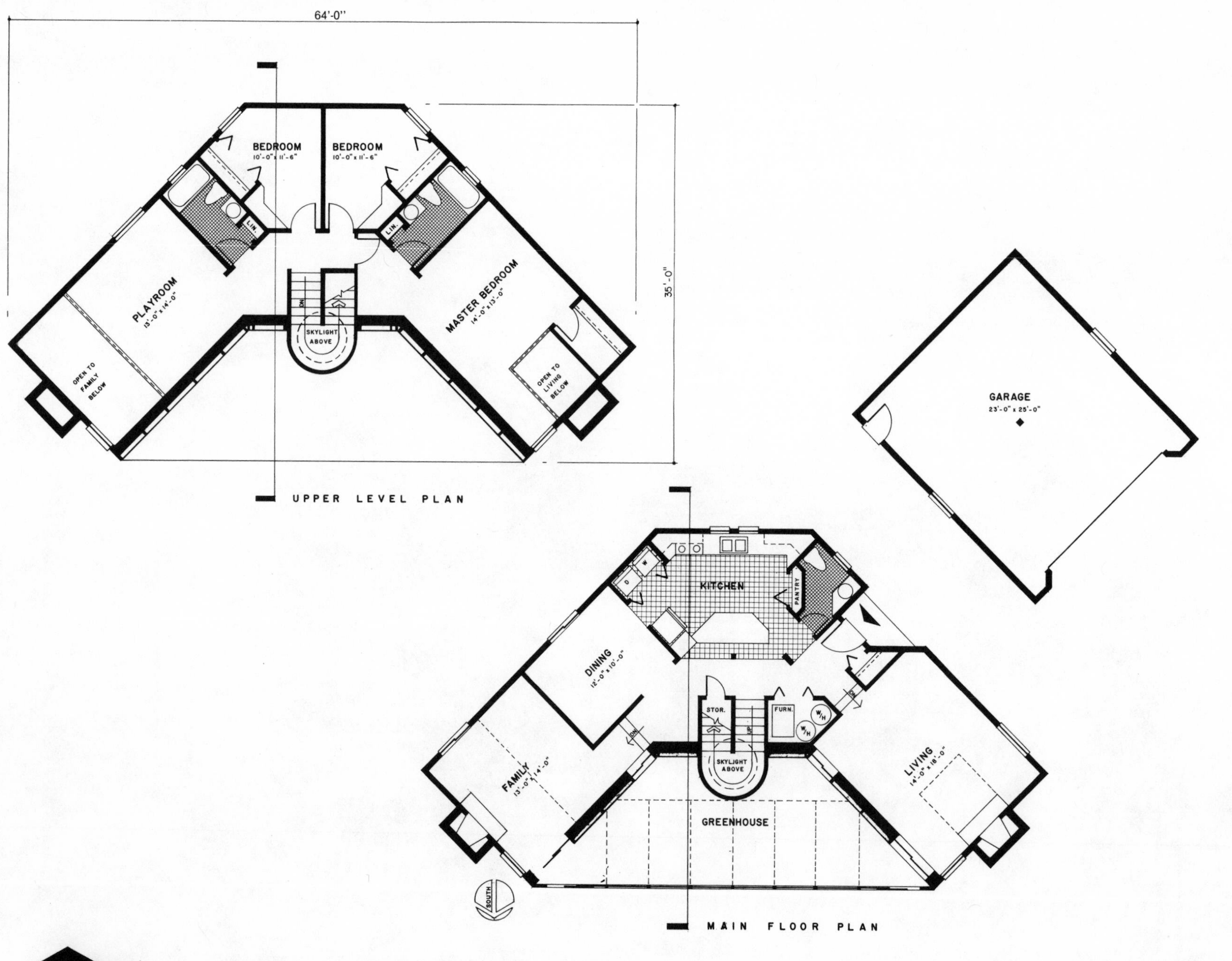

Space Time Designs inc.

S/T 10 - GREENHOUSE GALLERY

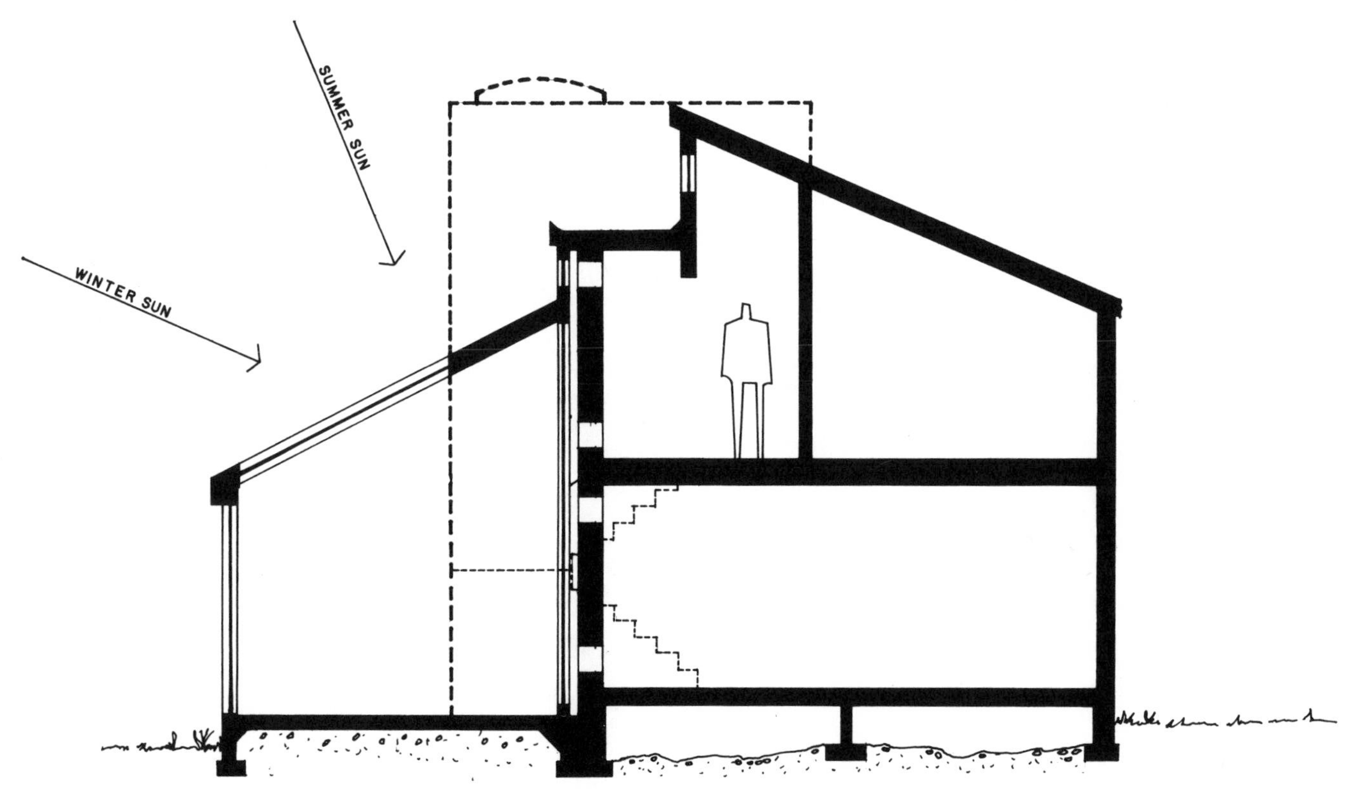

SECTION

SOLAR TRI-LEVEL
S/T-11

This contemporary home, incorporating the dynamics of exterior diagonals with a traditional tri-level floor plan, maximizes the potential active collector area without compromising living spaces. Designed to blend well into any contemporary neighborhood, it features a sunken living room with cathedral ceiling, a spacious dining room and a cozy family area five steps below the compact kitchen and eating nook. The lower level powder room and upper level skylit planter are encircled by a split stairway which opens out to a gallery hall overlooking the living room.

With the main living area opening directly to the backyard, this home allows for close child supervision and easy access to a multitude of outdoor activities. The enclosed family room greenhouse, which can be illuminated at night, will provide aesthetic as well as actual warmth to this space.

A steep north roof slope, absence of north windows, and heavily insulated walls should allow this home maximum solar efficiency.

Main floor — 815 sq. ft.
Lower floor — 195 sq. ft.
Upper floor — 980 sq. ft.
Greenhouse volume — 320 cu. ft.
Total square footage — 1,990

Package price — $390.00 — reverse plan add $100
3 bedrooms plus playroom
2½ bathrooms

Passive Features
Greenhouse glazing — 60 sq. ft.
Potential mass storage — 100 cu. ft.

Active Features
Potential collector area — 850 sq. ft.
Potential storage volume — 448 cu. ft.
Collector tilt — 60°
Water pre-heat system

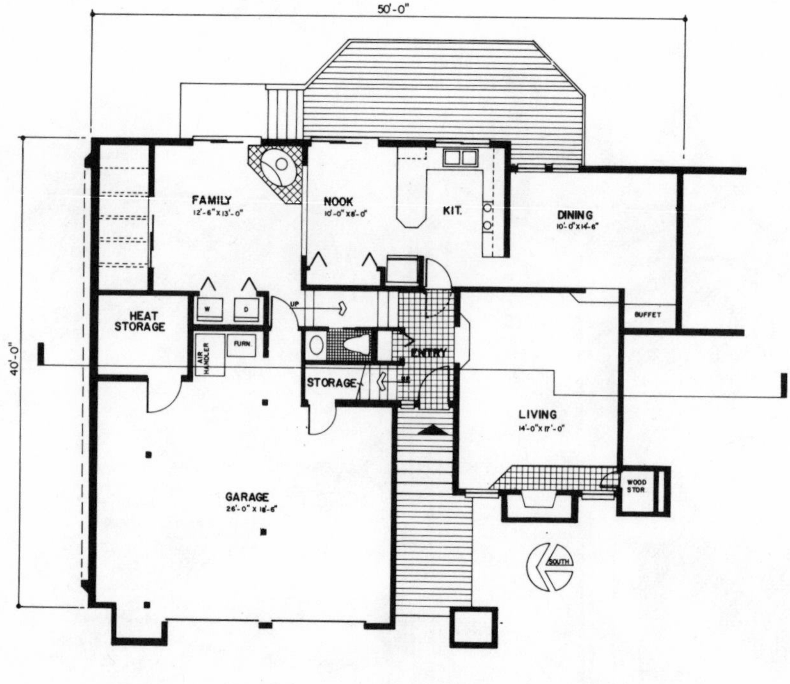

MAIN FLOOR PLAN

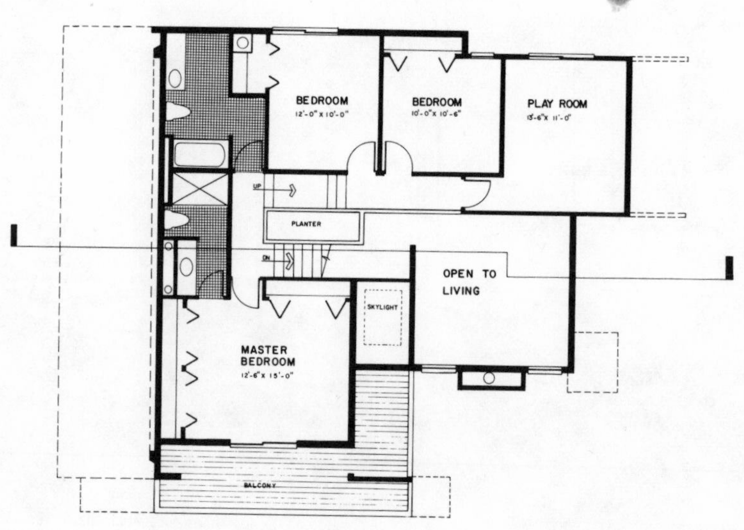

UPPER LEVEL PLAN

S/T 11 - SOLAR TRI-LEVEL

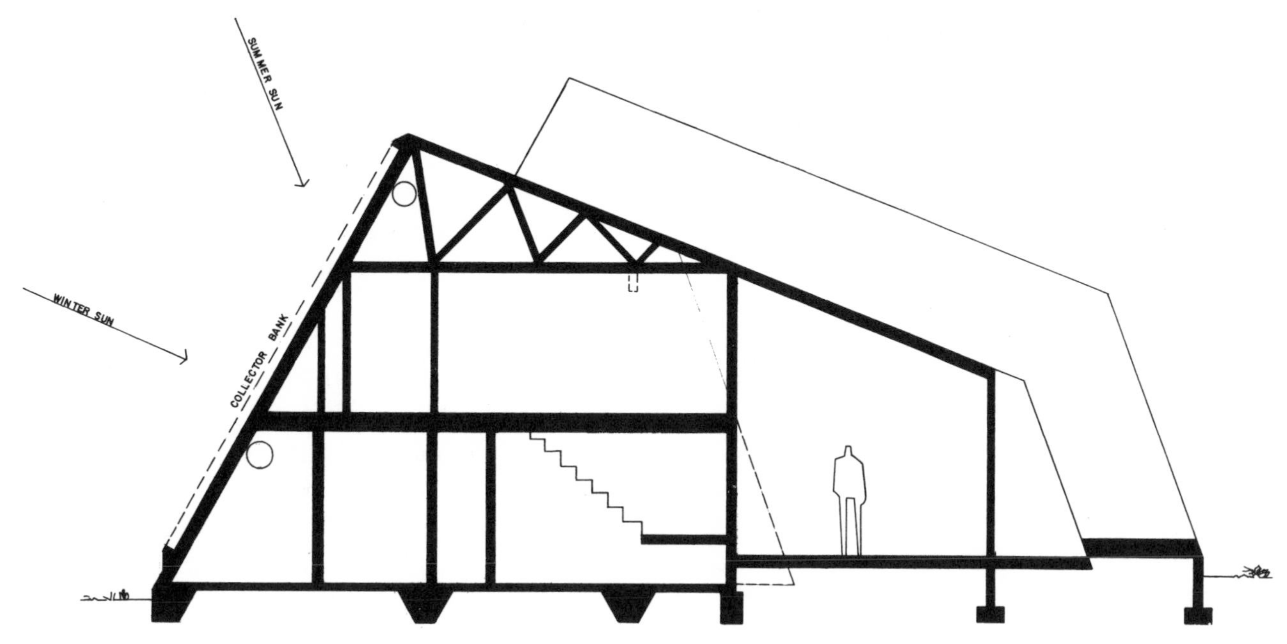

SECTION

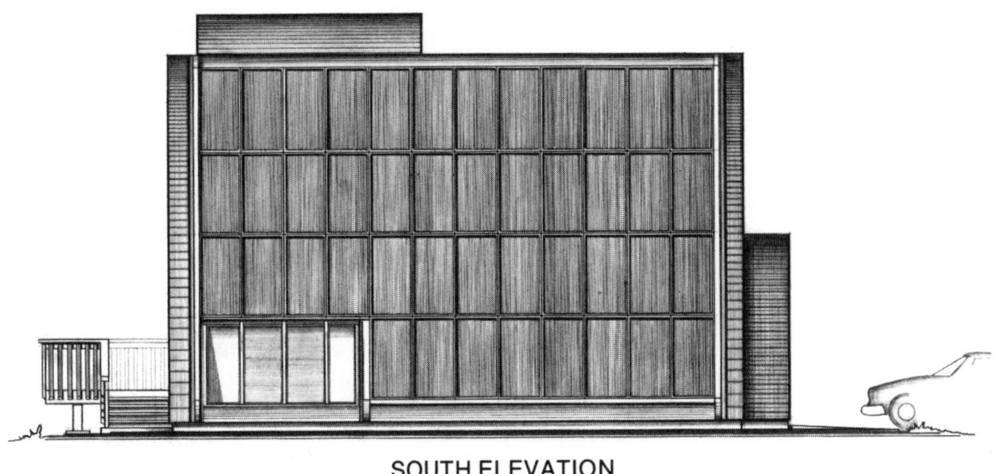

SOUTH ELEVATION

SPACEDOME I
S/T-33

This two-story, underground hemispherical dome is the theoretical optimum form for energy conservation, economy of structural materials, and thermal circulation. Adopting the most natural form in the universe and reminiscent of native American traditional buildings, this home is as valid an energy conserver as an Eskimo igloo or a Hopi desert hogan . . . and it allows maximum preservation of your natural landscape.

One enters at ground level through a double air lock entry to an open circular staircase. Graceful assymetrical shapes abound in the inner spaces of this home while still allowing for standard furniture arrangements. Conveniently located off the living room is a spatially interesting, pie-shaped master suite with access to kitchen/eating areas for midnight snacks. Upstairs are two additional sleeping rooms with private study and open play area. Both open to a balcony above the lofty entry. A central, 7-foot-diameter circular lightwell tracks the sun across the upper floor and stairwell. A creative builder should be able to construct this dome at a cost close to standard construction costs.

Forming design and complete engineering must be calculated and supervised by local engineers. Given sufficient demand, Space/Time Designs will be developing a universally-adaptable engineering and forming system. Please contact us at (602) 282-3639 for further information. For now, reproduceable structural drawings are included in the package for the use of your local engineer.

Main floor — 1,365 sq. ft.
Upper floor — 1,090
Total square footage — 2,455

Package Price — $780.00
3 bedrooms plus study
2½ bathrooms
30′ diameter cardome plans included in package price

Passive Features
Approximately 90% earth sheltered
South-facing glass — 68 sq. ft.
Structural mass — 945 cu. ft. concrete shell

Active Features
Water pre-heat system mounted in earth above

PETERSON '80

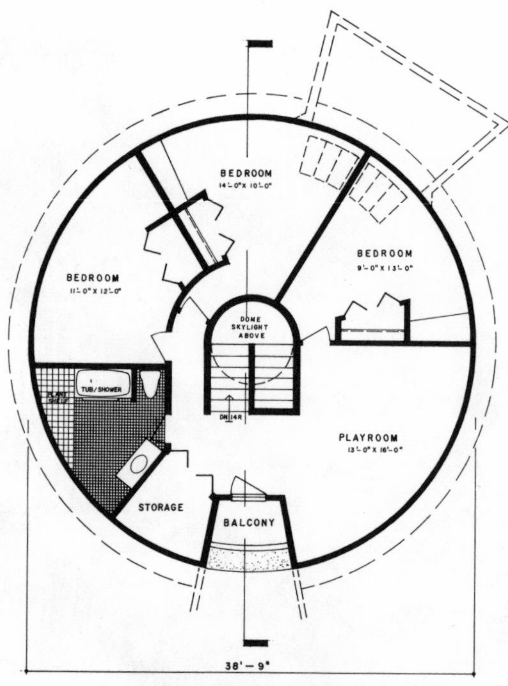

UPPER FLOOR PLAN

MASTER SUITE
14'-0" X 18'-0"
TUB/SHOWER
NOOK
SNACK BAR
KITCHEN
PANTRY
W. I. CL.
FAMILY ROOM
16'-0" X 14'-0"
LIVING ROOM
15'-0" X 16'-0"
FOYER
ENTRY
W. D.
42'-0"

LOWER FLOOR PLAN

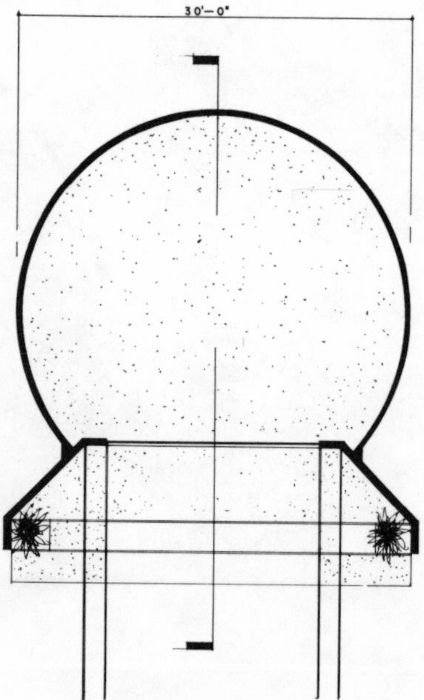

GARAGE PLAN

Space Time Designs inc.

S/T 33 - SPACEDOME I

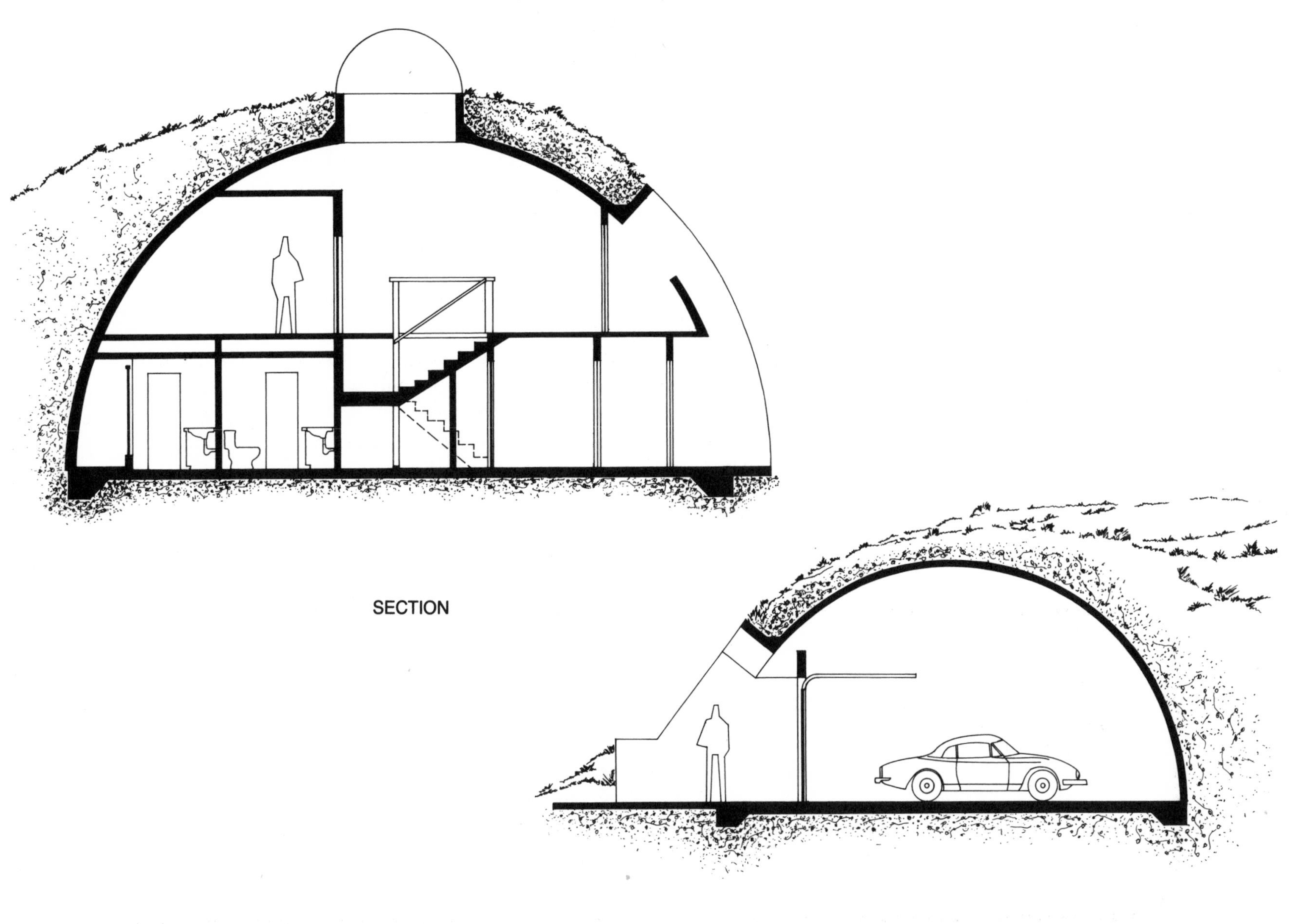
SECTION
GARAGE SECTION

SUNTREK
S/T-13

Suntrek is total experience, conceived by Dave Fordon in the spirit of tomorrow. The spatial dynamics and natural wonder of the interior atrium radiate into every area of the hovering design. The sunken entry explodes into the space hollowed out of the center of the house and gives access to a main level bedroom and to the sculptural steps leading up to the principle living areas and the secluded studio. Part way up the open U-shaped stairway is the main bath, and another few steps lead you into the balconied master suite.

A starship loft suspended over the main floor bedroom will be flooded with a cascade of light during the day and becomes a moonlit dream scene at night.

Its semi-pyramidal form gives this home an ideal start on energy conservation. The compact arrangement of customized collectors, air handler, heat storage, and water pre-heater should minimize the cost of mechanical installation. A decorator's dream, this home's capability for contributing to an incredible living experience is limited only by your imagination.

Plans available in metric dimensions upon request.

Main floor — 1,260 sq. ft.
Low headroom finished space — 112 sq. ft.
Upper floor — 466 sq. ft.
Total square footage — 1,838

Package price $390.00
2 bedrooms plus studio
2 bathrooms

Passive Features
Skylit atrium and entry

Active Features
Potential collector area — 220 sq. ft.
Potential storage volume — 240 cu. ft.
Collector tilt — 90°
Suggested water pre-heat system
is an air-to-water heat exchanger

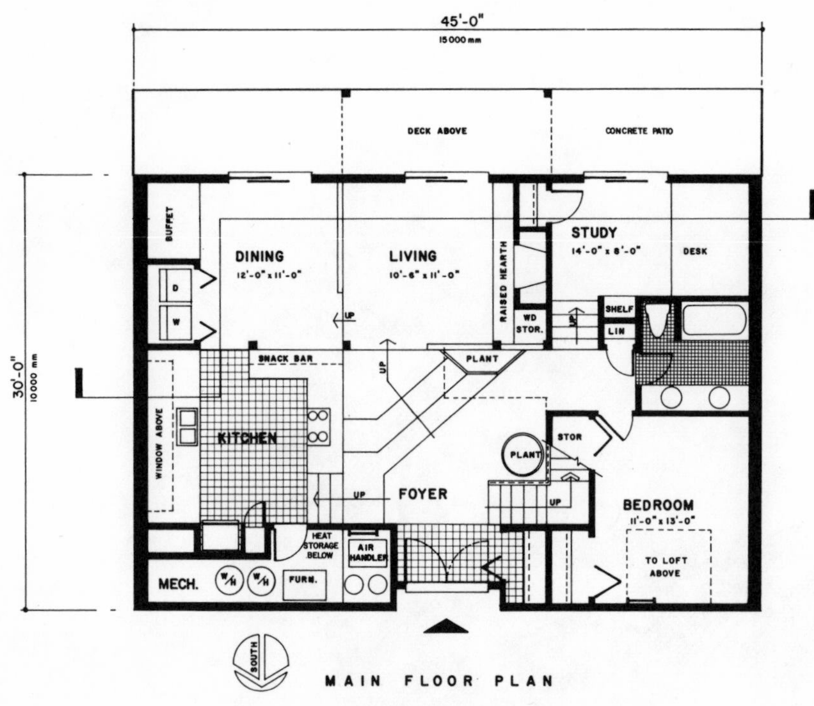

MAIN FLOOR PLAN

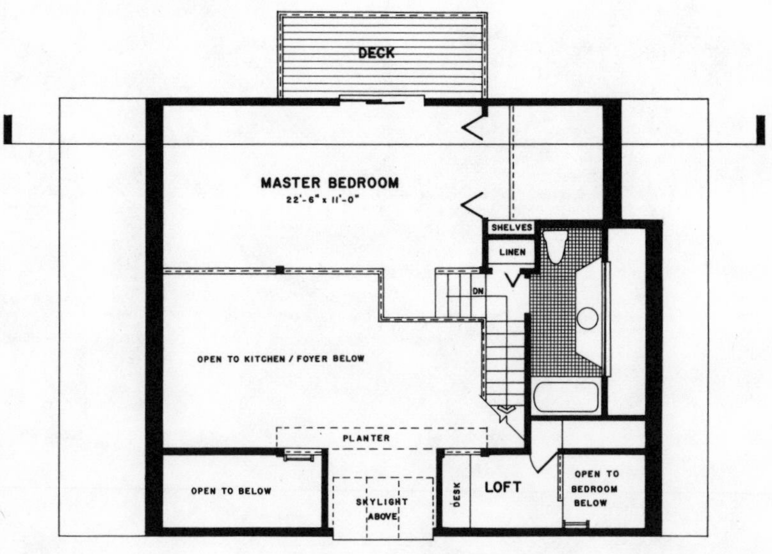

UPPER LEVEL PLAN

Space Time Designs inc.

S/T 13 - SUNTREK

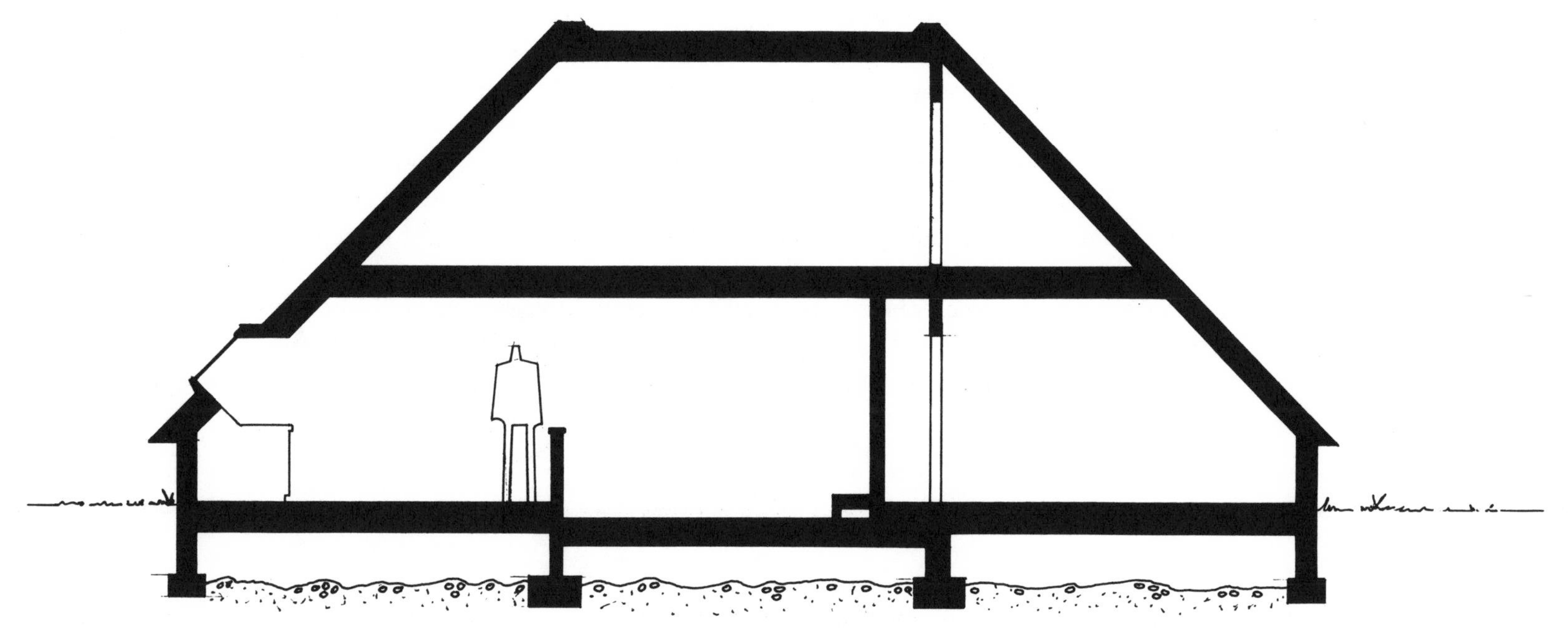

SECTION

TAHOE
S/T-34

This house is designed for a south-facing slope with street access and main entry on the north, l allows nearly 50% earth sheltering at the lower level. The focal point of the house is the tral deck area of the upper solarium, surrounded by the main living spaces which partake of ght and exciting structural complexity. The airlock entry at the second floor level brings you into either the warm bright spaces of the living room, with its native stone fireplace and floor-to-ceiling bookcase wall, or into the large kitchen featuring a spacious pantry with "California cooler," and a central island. The adjacent dining area is bathed in light on two sides from the greenhouse and overhead from skylights.

The bedrooms and study area on the lower floor face the sunny solarium, while closets, laundry and baths provide buffering to the north. Although exterior glazing on all but the south side is kept to a minimum, an ingenious arrangement of interior glass allows light to penetrate the entire structure. The location of stairs in the solarium conserves interior floor area and heightens the integration of activity between the solarium and adjacent interior spaces. Other design features included but not shown are the garage, covered walkway and entry deck. Structurally designed to support heavy snow loads, this design can also be modified for more economical building in a more temperate climate.

Main floor — 1,022 sq. ft.
Upper floor — 1,054 sq. ft.
Greenhouse volume — 9,400 cu. ft.
Total interior square footage — 2,176

Package Price — $550.00
4 bedrooms plus study
3½ bathrooms

Passive Features
Exterior vertical greenhouse glazing — 428 sq. ft.
Greenhouse 45° roof glazing — 480 sq. ft.
Geo-thermal air envelope system

Active Features
Suggested water pre-heat system

CASTILLO '79

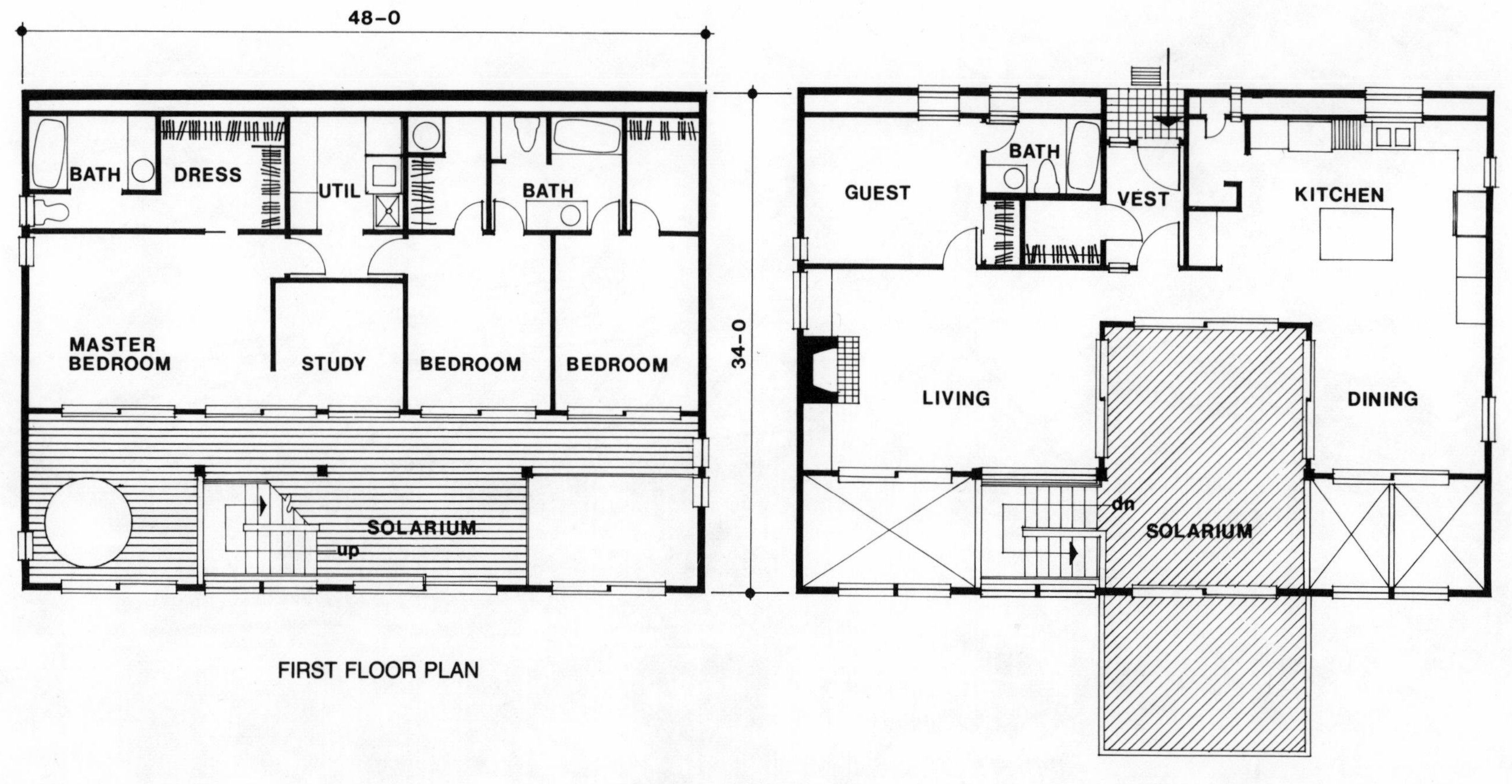

FIRST FLOOR PLAN

SECOND FLOOR PLAN

Space Time Designs inc.

S/T 34 - TAHOE

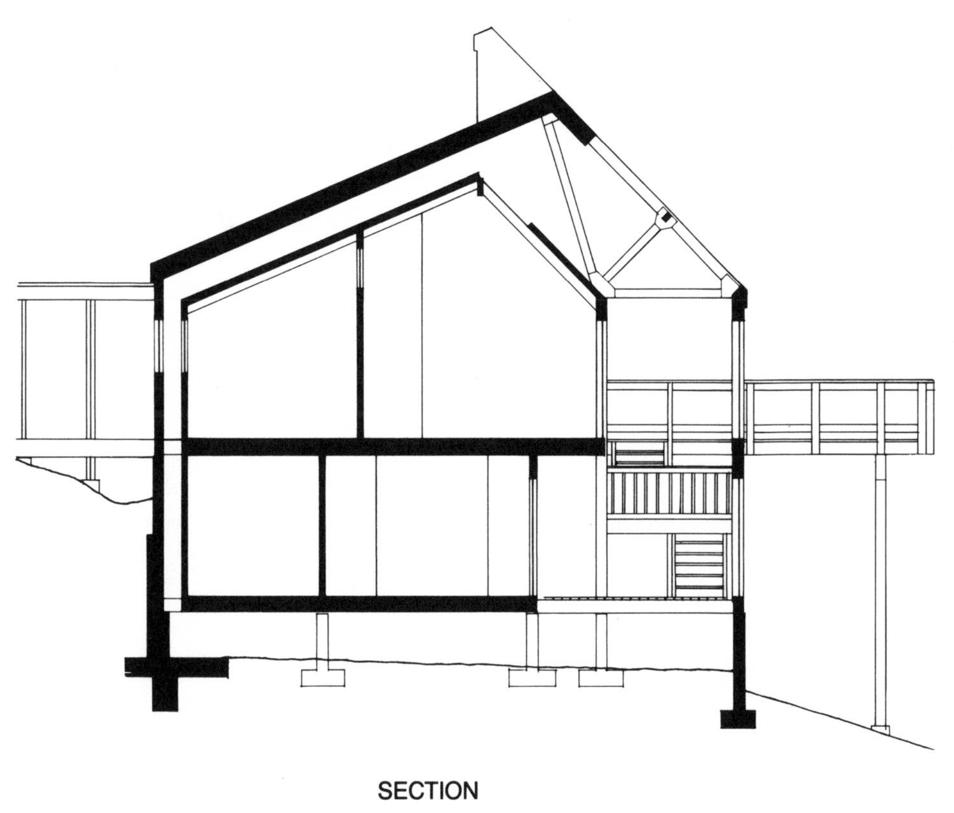

SECTION

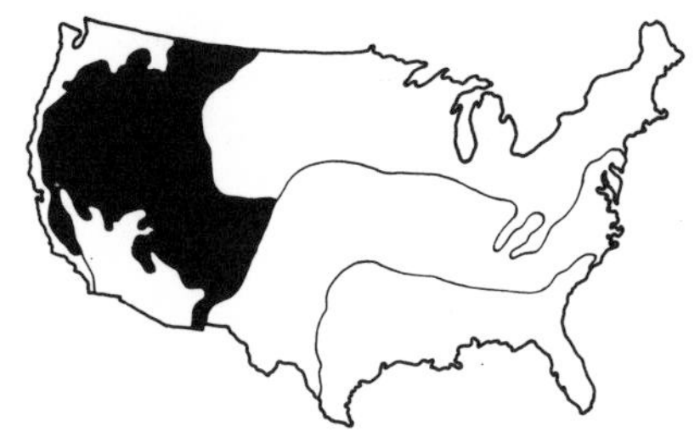

SKY BUBBLES
S/T-15

Have a house site with limited sun exposure? Want your house surrounded by trees? Try building this contemporary sculptural home with its uniquely designed solar furnace garage. Offering a wealth of storage, workshop, or studio space, the garage's south wall or south roof line (depending on its orientation) can supply your home with a surprising amount of solar heat. Careful engineering can allow stored heat to be drawn from the collector or storage area through a super-insulated and controlled underground duct system for distribution by a forced air furnace.*

The home itself is designed especially to compliment the contemporary shed roof garage. Inside the airlock foyer, a quarry tile entry leads to an open living area. The horseshoe-shaped dining area, beyond a sunken living room, provides ambiance around a round dining table. A compact kitchen with handy laundry facilities is open to a cozy family room. Up the round skylit stairway are two small bedrooms and a lovely master suite with its own naturally-lit den and hall. The sun deck completes the master suite.

Although not a solar structure, minimum volume, super-insulated walls, an airlock entry, and a central masonry fireplace will make this home adequately energy efficient.

*The garage plans are ideally suited to be adapted to other shed roof style structures, and can be purchased separately with specifications for alternate door and collector locations. If you diagram your desired siting needs with your order, you will be assured of the correct plan.

Main floor — 1,066 sq. ft.
Upper floor — 886 sq. ft.
Total square footage — 1,952

Package price — $520.00 including garage
Garage only (four sets) — $170.00
3 bedrooms plus den
2½ bathrooms

Active Features
Potential collector area — 567 sq. ft.
Potential storage volume — 512 cu. ft.
Collector tilt — 90° or 30°

Sisko

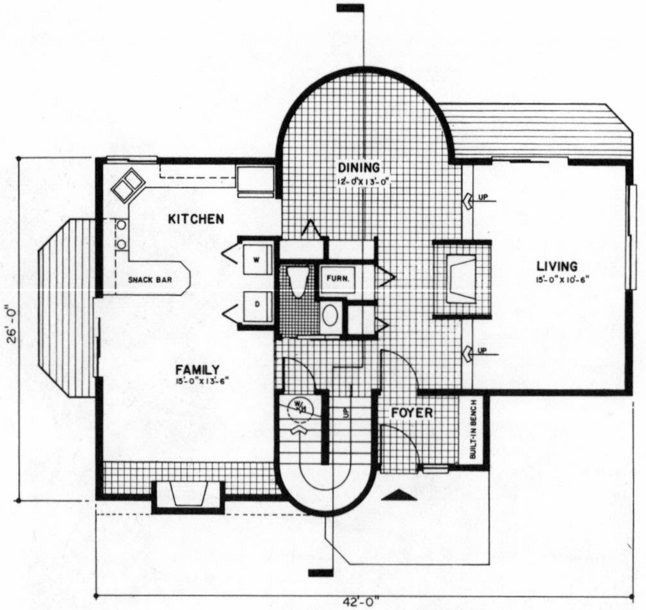

DINING
12'-0" x 13'-0"

KITCHEN

SNACK BAR

W

D

FURN.

UP

LIVING
15'-0" x 10'-6"

UP

FAMILY
15'-0" x 13'-6"

W/H

UP

FOYER

BUILT-IN BENCH

26'-0"

42'-0"

MAIN FLOOR PLAN

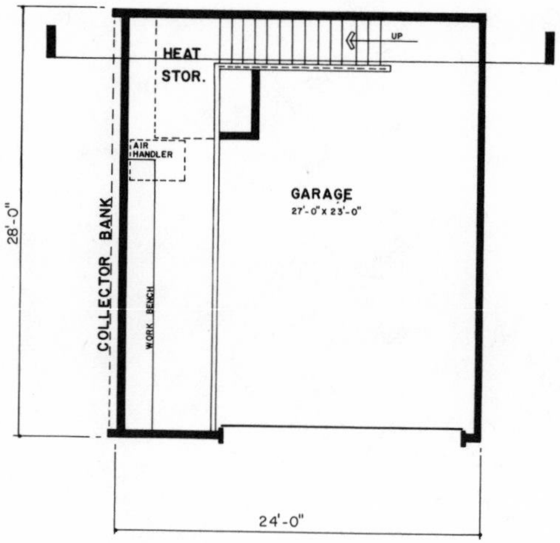

UP

HEAT
STOR.

AIR
HANDLER

COLLECTOR BANK

WORK BENCH

GARAGE
27'-0" x 23'-0"

28'-0"

24'-0"

GARAGE FLOOR PLAN

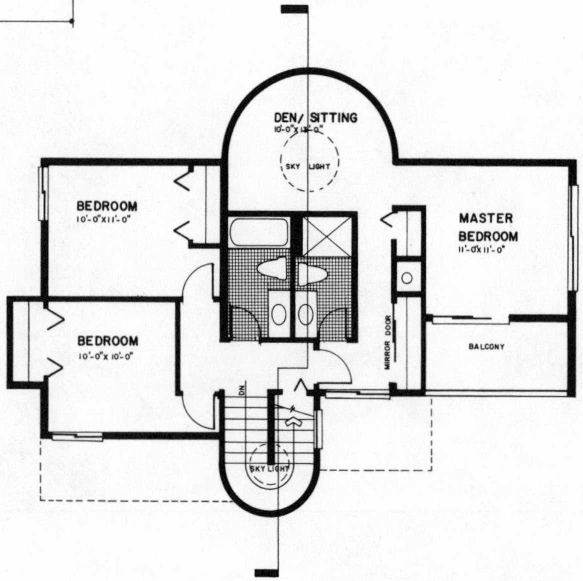

DEN/SITTING
10'-0" x 12'-0"

SKY LIGHT

BEDROOM
10'-0" x 11'-0"

MASTER
BEDROOM
11'-0" x 11'-0"

BEDROOM
10'-0" x 10'-0"

MIRROR DOOR

BALCONY

DN

SKY LIGHT

UPPER LEVEL PLAN

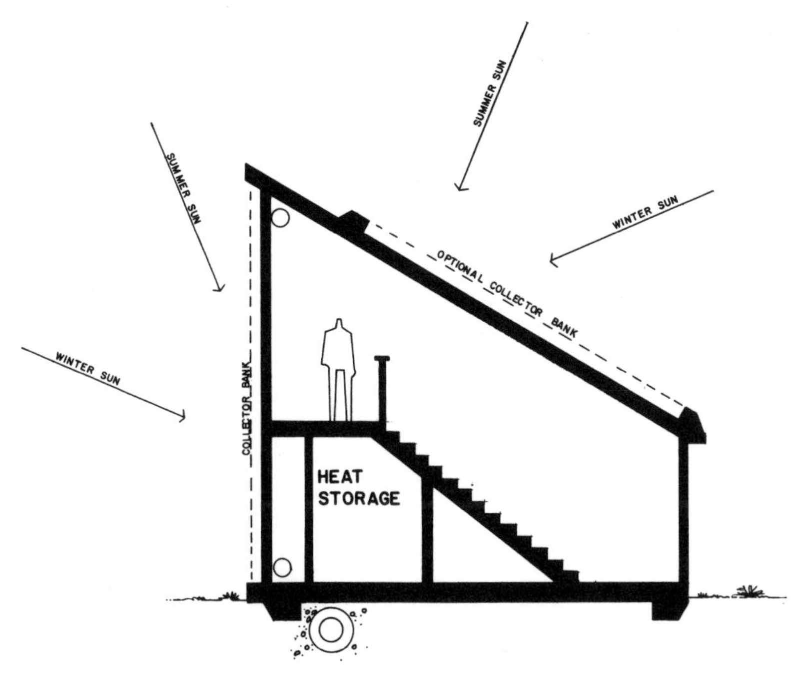

GARAGE SECTION

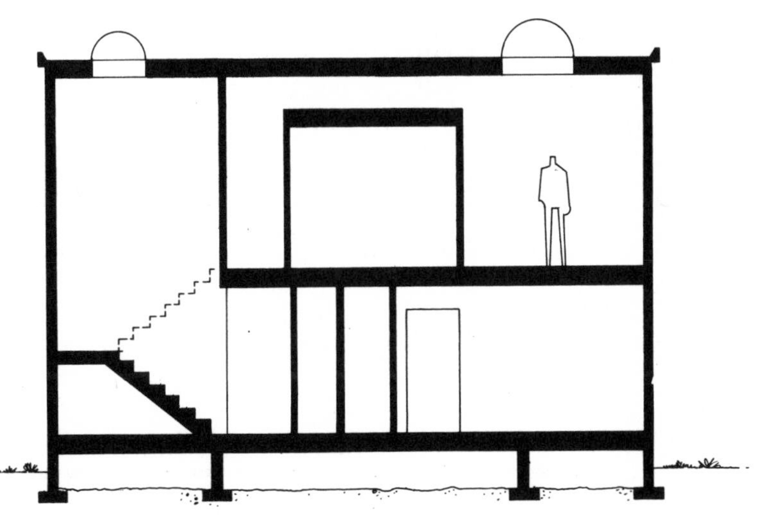

SECTION

OLD GREEN THUMBS
S/T-16

With active/passive capability of about equal potential, this home's storage area can be engineered to gather excess heat from the greenhouse as well as from the vertically mounted collectors. The close proximity of collectors, greenhouse, and water pre-heater minimizes the complexities of heat storage and dispersal. Warmth from the greenhouse can enter directly into the kitchen, dining area, and living room. Windows along the south wall of the living room allow for direct gain heating, and enable you to enjoy the view and the fire at the same time.

A private retreat, the third floor master bedroom has its own separate skylit study open to the living room below. Mirrored bypass closet doors add both light and psychological space to the bedroom itself.

In some ways as new as tomorrow, the basic design of *Old Green Thumbs* is comfortably traditional. Finishing the billiard room space on the ground floor as two additional bedrooms (be sure to describe your intentions with your order)* would make this house ideal for a large conservation-minded family.

Also included with plans is garage as shown in rendering.

*For fourth and fifth bedrooms use order number S/T-16A.

Main floor — 900 sq. ft.
Lower floor — 952 sq. ft.
Upper floor — 543 sq. ft.
Greenhouse volume — 2,975 cu. ft.
Total square footage — 2,395

Package price — $620.00
3 bedrooms plus study
2½ bathrooms

Passive Features
Direct gain glazing — 80 sq. ft.
Greenhouse glazing — 506 sq. ft.
Potential mass storage — 500 cu. ft.

Active Features
Potential collector area — 253 sq. ft.
Potential storage volume — 320 cu. ft.
Collector tilt — 90°
Water pre-heat system

SOLTNER

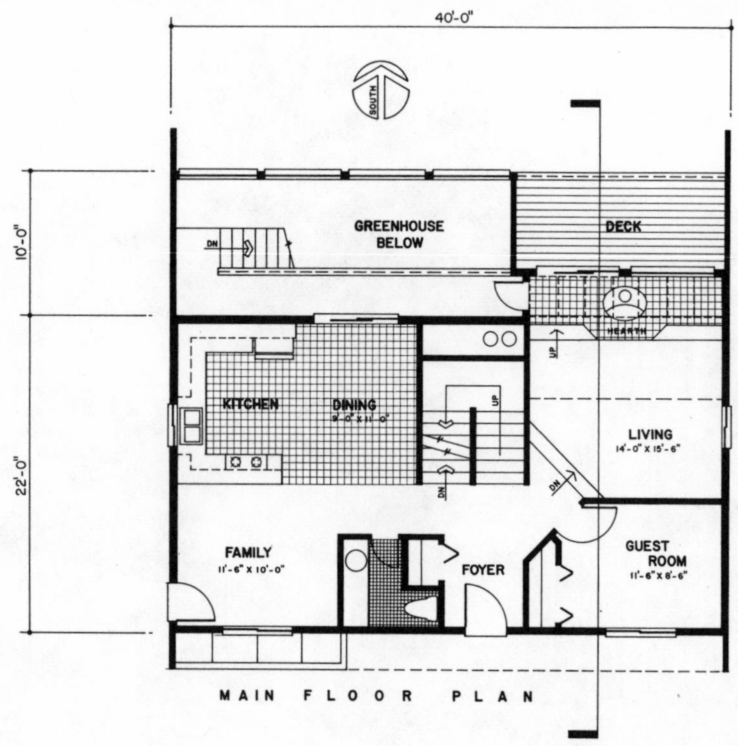

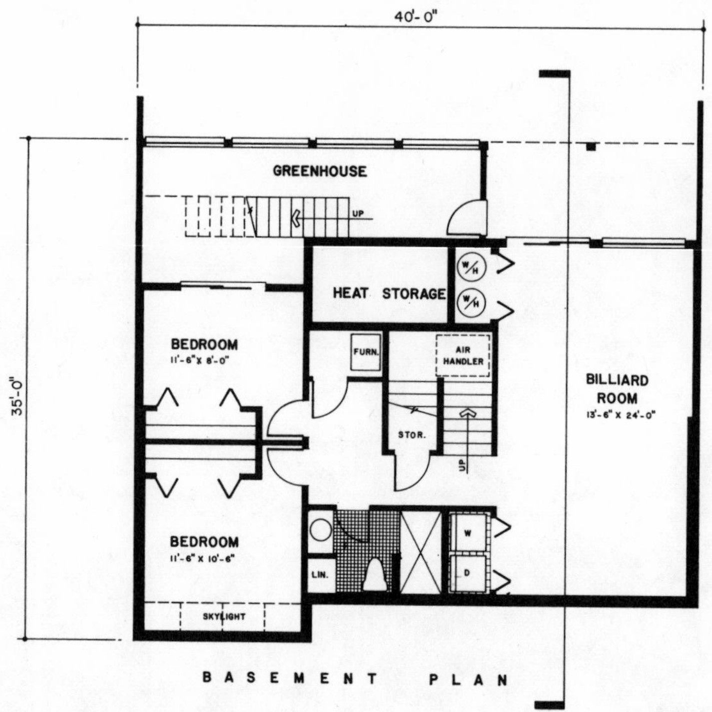

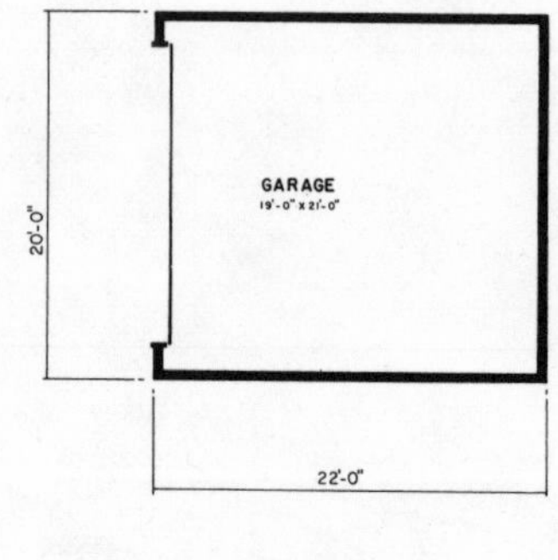

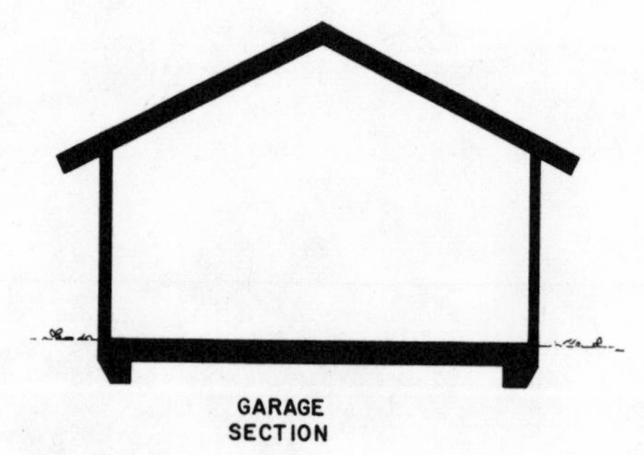

Space Time Designs inc.

S/T 16 - OLD GREEN THUMBS

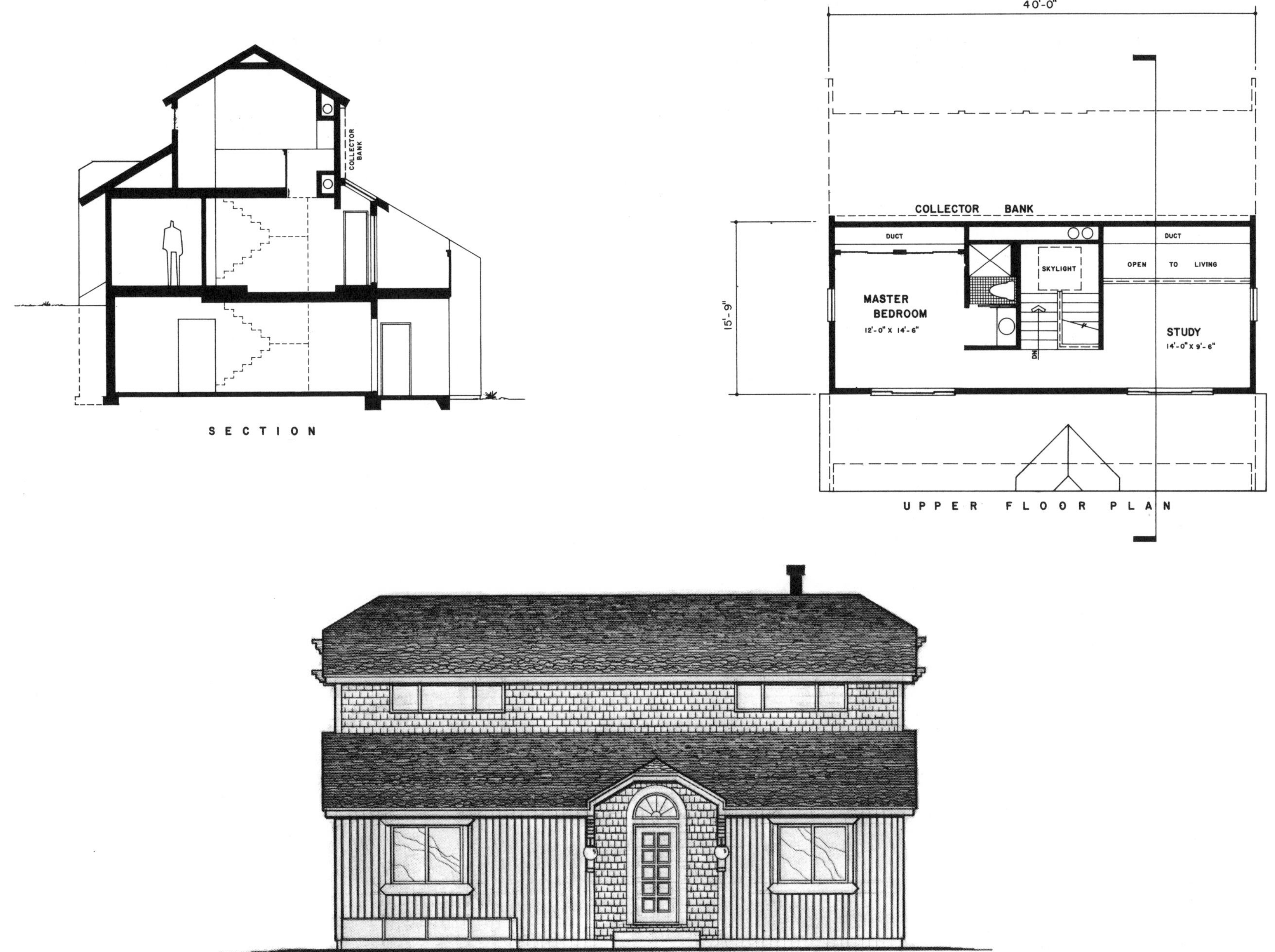

SECTION

UPPER FLOOR PLAN

NORTH ELEVATION

THE CONVECTION CONNECTION
S/T-17

This home, with careful sizing of mass and with provision for insulating and shading devices, can be fine-tuned to provide sociable living with lots of solar comfort. Optimally proportioned for energy efficiency, this house will collect and circulate warm air using only the natural laws of heat transfer.

The mirror-like mass wall is broken only by colorful stained glass illuminating the entry. The greenhouse adjoining the kitchen passively heats this area and provides fresh vegetables as well. Mass wall vents and high northern windows provide summer ventilation over the 8-foot-high master suite walls. The restrained design of the house will accommodate a variety of home furnishing styles.

Practical features include a generous shop area in the garage, a main floor den or study, small entry area off a large and convenient island kitchen, and a ground floor guest bath. Upstairs, a den, sun deck, large walk-in closet and a skylit two-lavatory vanity provides the master suite with its own luxury.

Main floor — 1,019 sq. ft.
Upper floor — 1,014 sq. ft.
Greenhouse volume — 1,760 cu. ft.
Total square footage — 2,033

Package price — $370.00
3 bedrooms plus study plus den
2½ bathrooms

Passive Features
Greenhouse glazing — 232 sq. ft.
Mass wall glazing — 512 sq. ft.
Potential mass wall storage — 490 cu. ft.
Potential greenhouse mass — 400 cu. ft.

Active Features
Water pre-heat system
High cold air return for crawl space rock bin storage

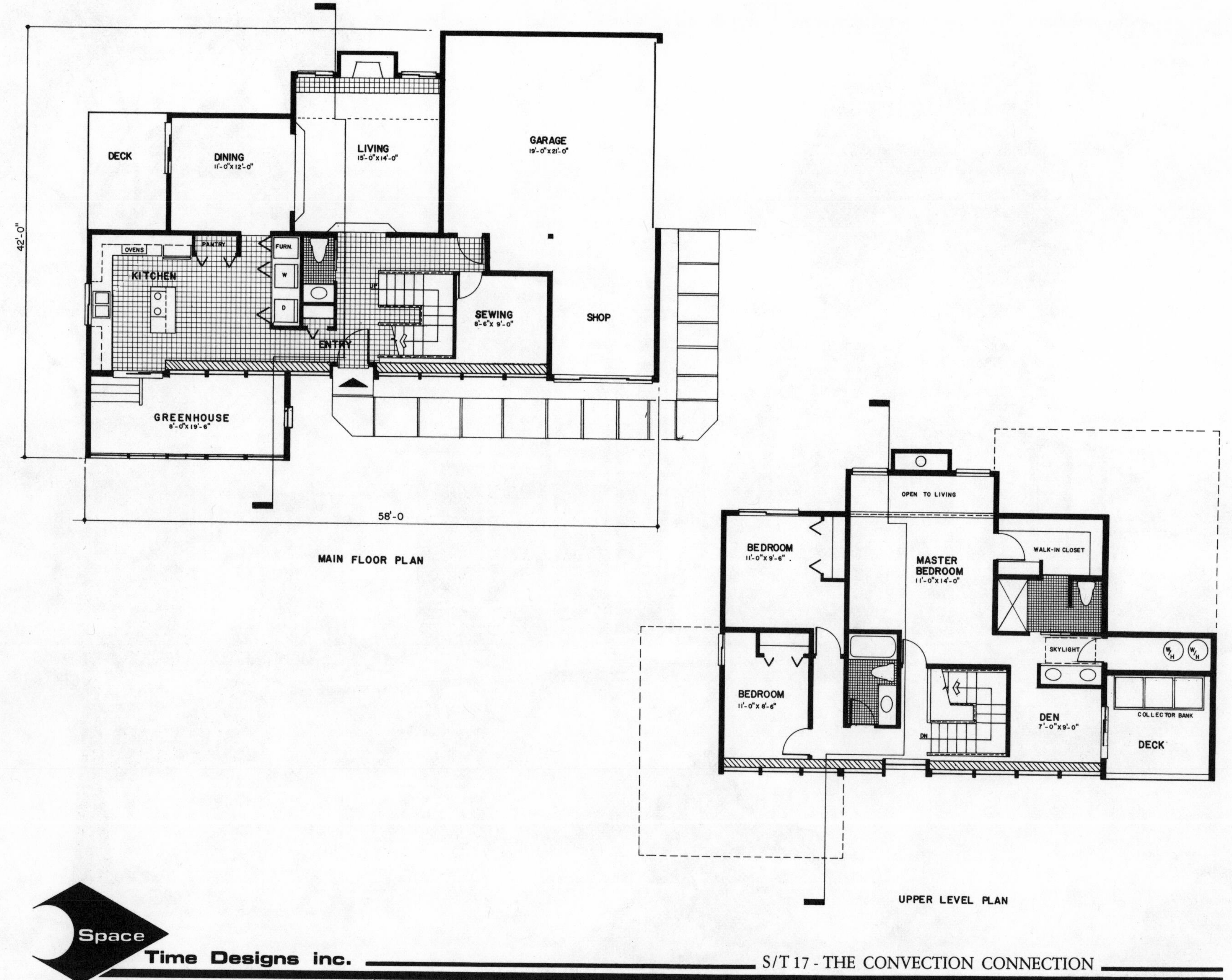

Space Time Designs inc.

S/T 17 - THE CONVECTION CONNECTION

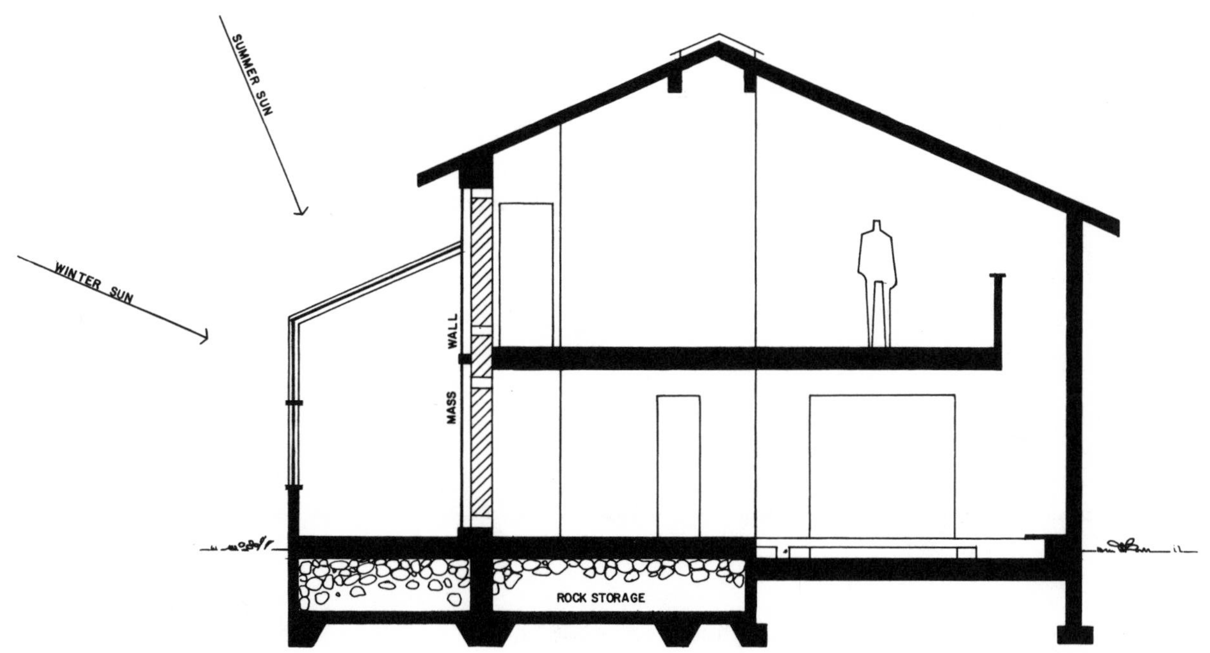

SECTION

INNERPEACE
S/T-18

The subtly curved lines of the north side of this design provide a sculptural contrast to the strong diagonal planes of the solar collectors and two-story greenhouse on the south. Uniquely zoned for privacy and convenience, the curved interior spaces suggest peace and tranquility.

Coming in the lower-level greenhouse entry, you may either go directly to one of the two spacious downstairs bedrooms which have room for play or study as well as for sleeping, or you may go topside to the main living areas and huge master suite. An intimate greenhouse living room with cheerful corner fireplace is open to the sculptural dining room. A spiral staircase to the exciting recreation area on the third level visually separates dining area and kitchen.

The freestanding fireplace in the master bedroom sitting area warms this space and provides a cozy place to curl up with a good book, while the high, south-facing skylights flanking the fireplace admit light and warmth without sacrificing privacy.

Heavy insulation in the 2'' x 6'' stud walls, minimum north windows, and warmth from the greenhouse and two fireplaces should combine with the active systems to give this plan high heat efficiency; and with ample working space in the greenhouse and the huge roof garden area,* this design could be a gardener's dream.

*Not recommended for climates with heavy snow, the plan can be ordered with an alternate raised flat roof which will not trap snow build-up. Please specify which you prefer when ordering.

Main floor — 1,444 sq. ft.
Lower floor — 826 sq. ft.
Upper floor — 575 sq. ft.
Greenhouse volume — 2,520 cu. ft.
Total square footage — 2,845

Package price — $530.00
3 bedrooms
3 bathrooms

Passive Features
Greenhouse glazing — 240 sq. ft.
Potential mass storage — 700 cu. ft.

Active Features
Potential collector area — 351 sq. ft.
Potential storage volume — 350 cu. ft.
Collector tilt — 45°
Suggested water pre-heat system is an air-to-water heat exchanger

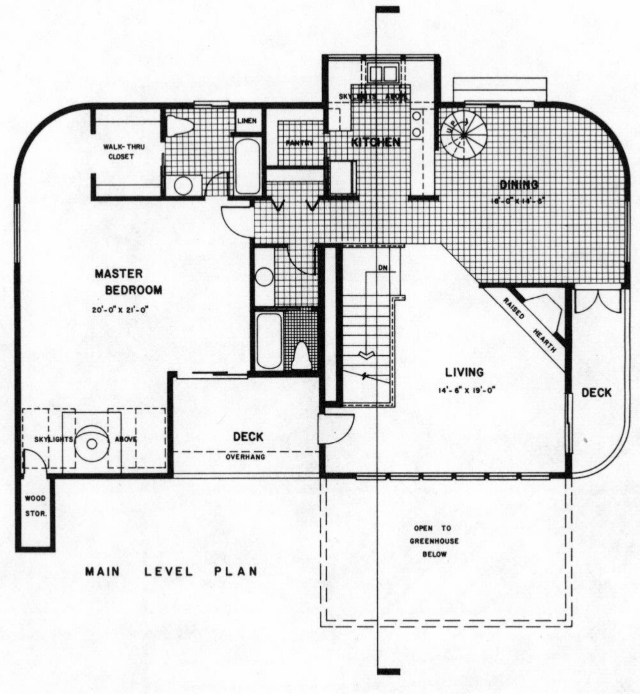

MAIN LEVEL PLAN

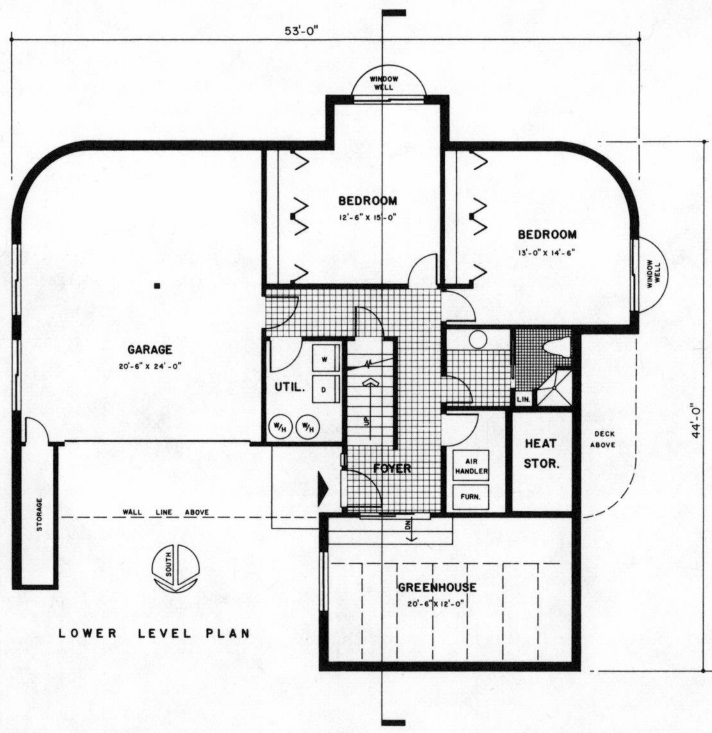

LOWER LEVEL PLAN

Space Time Designs inc.

S/T 18 - INNERPEACE

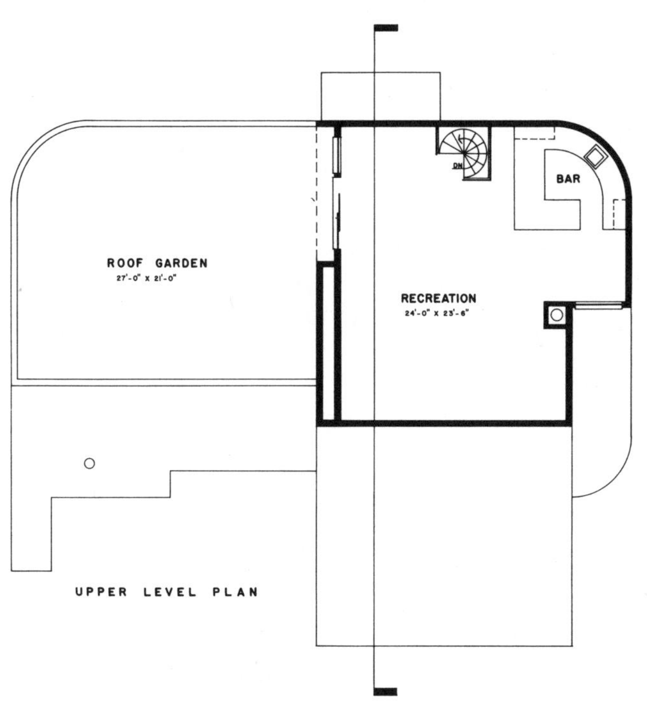

UPPER LEVEL PLAN

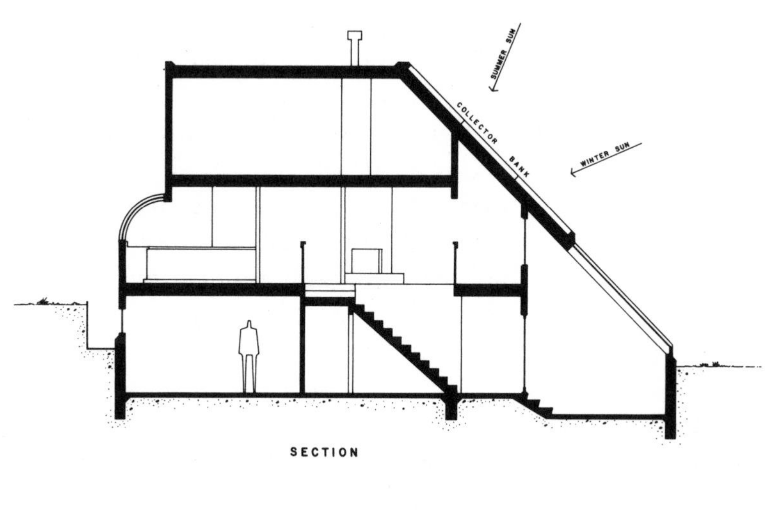

SECTION

BEHIND CLOSED DOORS
S/T-19

With the whole house acting as a direct gain heat trap, this design represents the ultimate in simplicity for both construction and operation. Ideal for a view site in a good sun zone, the home absorbs solar energy directly into its insulated mass floor, stunning stairwell wrapped with water-filled planters, and mass-laden balcony, and radiates the collected heat back into the conditioned spaces at night and in cloudy weather. During these latter periods, the two-story insulated folding wall is activated to preserve comfortable interior temperatures. East-west ventilation, south roof overhang (adjusted for local sun angles) and potential rooftop ponding permit summer cooling.

A compact, convenient floor plan maximizes construction simplicity without compromising livability. All the main living spaces radiate openly from the central heat-collecting stairwell, and provide a feeling of great space and exciting interior vistas with minimum square footage. With funky exposed post and beam construction, the visual warmth of this house should equal its actual warmth. If you combine super insulation with all these passive solar features, you may never have to resort to the back-up heating system located in the single-car garage.

Main floor — 972 sq. ft.
Upper floor — 650 sq. ft.
Total square footage — 1,622

Package price — $380.00
3 bedrooms
2½ bathrooms

Passive Features
Direct gain glazing — 350 sq. ft.
Potential mass storage — 500 cu. ft.

Active Features
Water pre-heat system

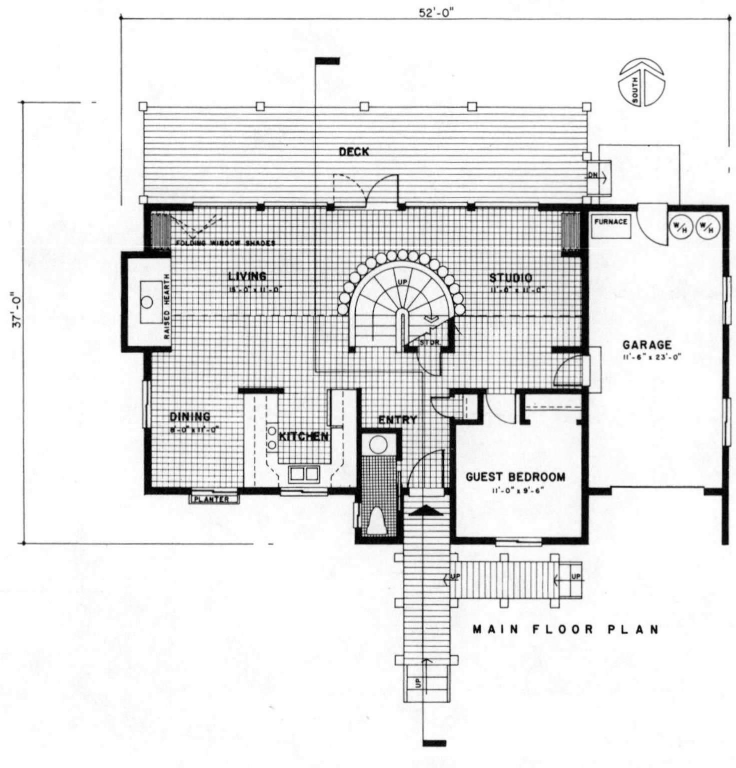

MAIN FLOOR PLAN

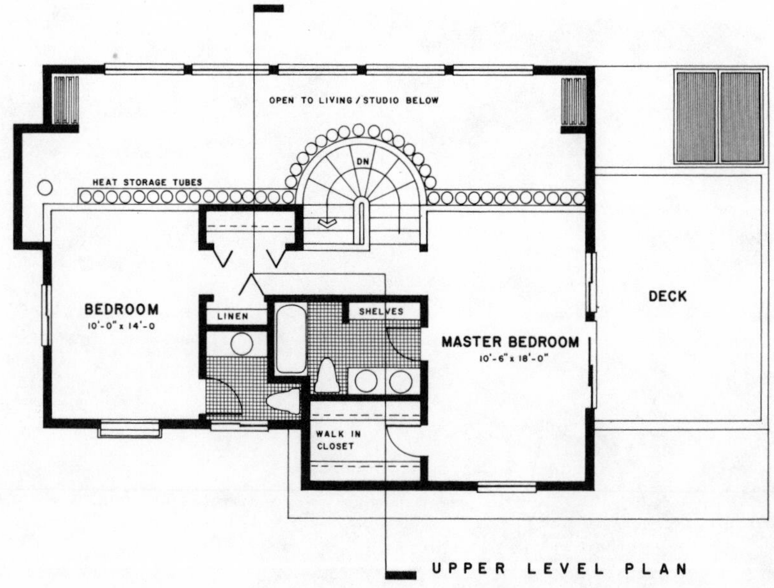

UPPER LEVEL PLAN

Space Time Designs inc.

S/T 19 - BEHIND CLOSED DOORS

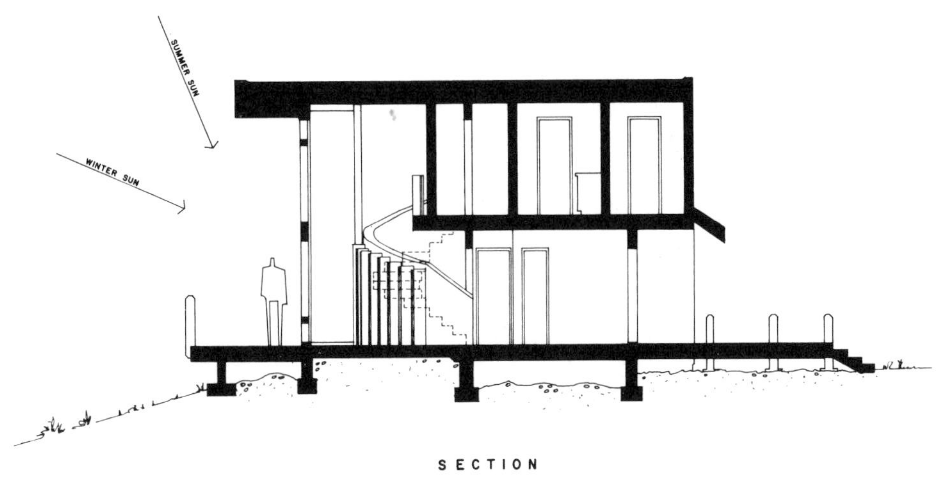

SECTION

NORTH ELEVATION

A TOUCH OF GREEN
S/T-20

This contemporary design with strong vertical lines and inclined planes is another active/passive hybrid. Both living room and kitchen have a "touch of green;" the mini-greenhouse beyond the kitchen sink provides an ideal place for supervising seedlings or convalescent plants.

From the entry flanked by the winding staircase, past the archway into the sunken living room, you come to a door which offers privacy to family activities centering around the triangular eating counter and the freestanding fireplace or wood stove in the generous family room area.

The north-side garage and nearly windowless north walls provide a barrier to cold outside temperatures. Maximum insulation in 2x6 24" o.c. perimeter wall framing and the centralized heat storage space should give this home high energy efficiency.

Flush-faced vertical siding (cedar or redwood) can accentuate this home's sculptural form and provide a foil for the vertical stained glass window illuminating the staircase. The overhead bubble gives additional light to the large second-floor planter.

A private den off the master bedroom is ideally suited for quiet evening reading or daytime sewing activities.

Main floor — 1,155 sq. ft.
Upper floor — 982 sq. ft.
Greenhouse volume — 720 cu. ft.
Total square footage — 2,137

Price of package — $540.00 —
reverse plan add $100
3 bedrooms plus den
2½ bathrooms

Passive Features
Direct gain glazing — 58 sq. ft.
Greenhouse glazing — 126 sq. ft.
Potential mass storage — 150 cu. ft.

Active Features
Potential collector area — 468 sq. ft.
Potential storage volume — 560 cu. ft.
Collector tilt — 60°
Water pre-heat system

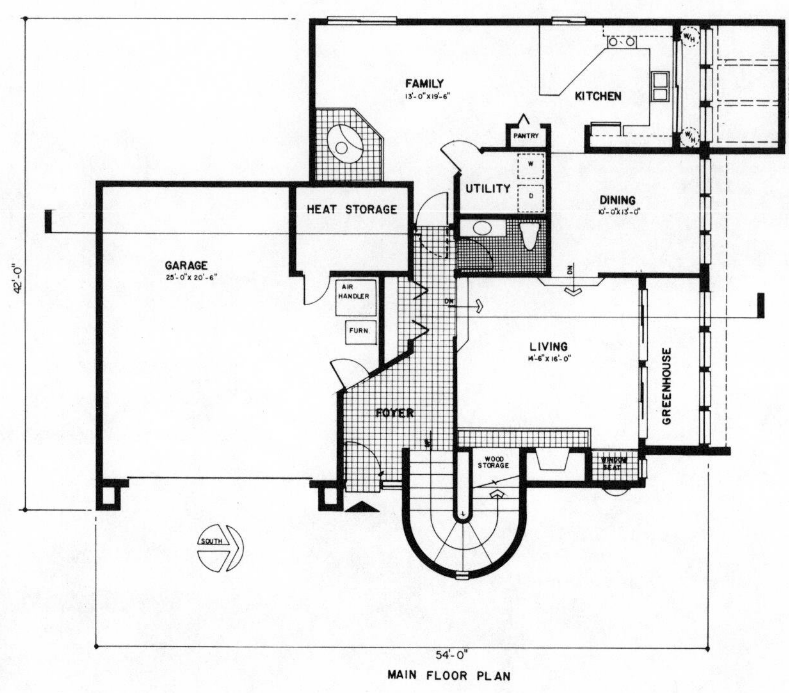

MAIN FLOOR PLAN

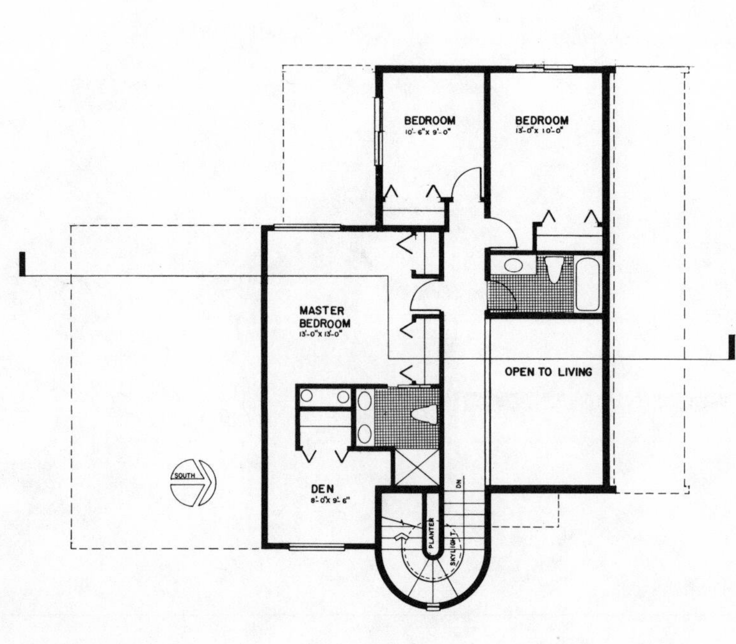

UPPER LEVEL PLAN

Space Time Designs inc.

S/T 20 - A TOUCH OF GREEN

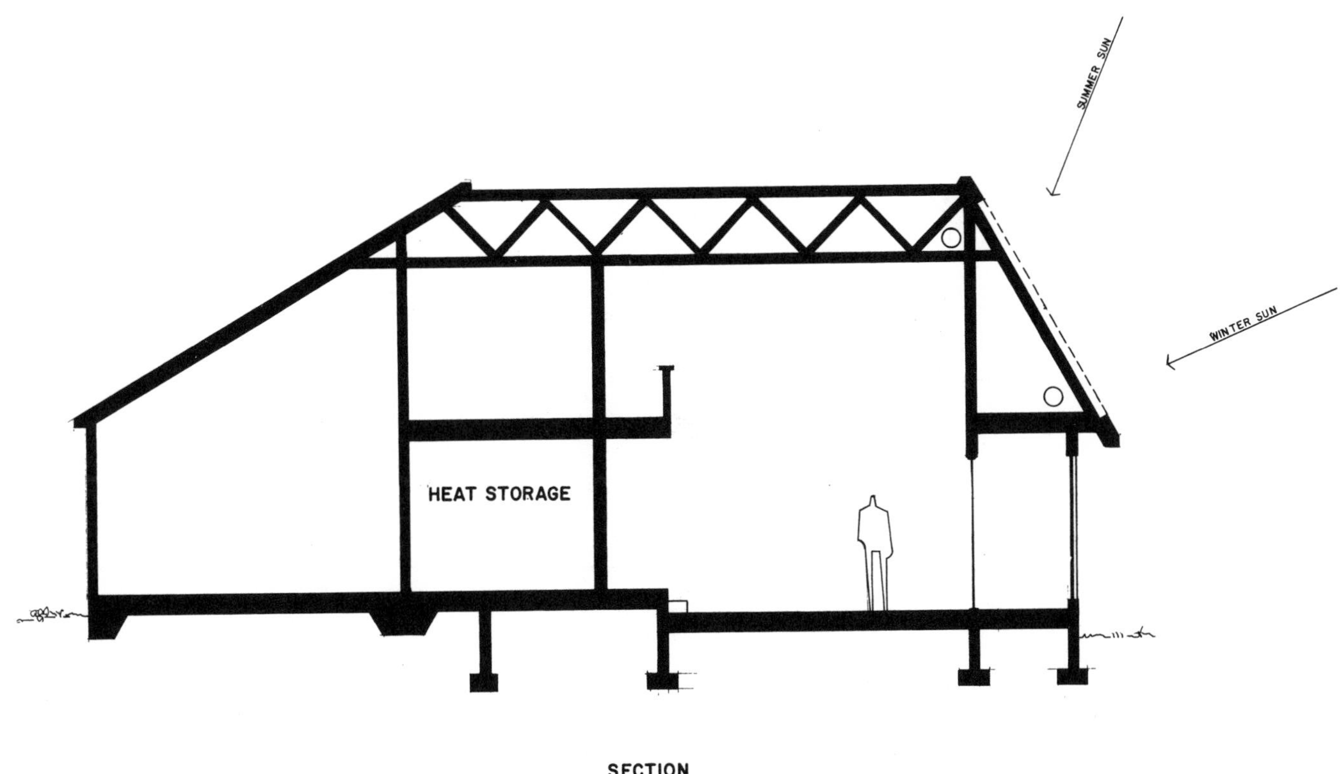

SECTION

EXPLORER
S/T-21

Step off the driveway and gaze through the space of round windows as if the end of a time tunnel were near. This three-story, thirty-five-foot tall uni-space structure contains over 700 square feet of active/passive solar aperture.

Beyond double entry doors, through a two-story entry and passing by the naturally lit pool-table-sized rec room (adjoining the guest bedroom and bath), climb into an enormous open living area. This lofty space is filled with light from a large skylight over the entry and two smaller ones in the spacious kitchen/family area. The living room features a cathedral ceiling and a fireplace flanked by stereo or book shelves. A light, open, and airy third floor sleeping loft could easily alternate with the main floor bedroom as the master suite. The design features both open areas for conviviality and secluded spaces for privacy.

Steep northern roof slope, underground first floor siting, ample greenhouse, thermal efficient fireplace on an interior wall, strategically placed heat exhausting north windows, and a ton of solar collectors should combine to make this one of the most effective solar designs in this book. Hot water pre-heating is possible through a heat exchanger in the air system.

Main floor — 1,452 sq. ft.
Lower floor — 798 sq. ft.
Upper floor — 482 sq. ft.
Greenhouse volume — 1,540 cu. ft.
Total square footage — 2,732

Package price — $720.00
3 bedrooms
3 bathrooms

Passive Features
Direct gain glazing — 78 sq. ft.
Greenhouse glazing — 108 sq. ft.
Potential mass storage — 300 cu. ft.

Active Features
Potential collector area — 525 sq. ft.
Potential storage volume — 360 cu. ft.
Collector tilt — 60°
Water pre-heat system

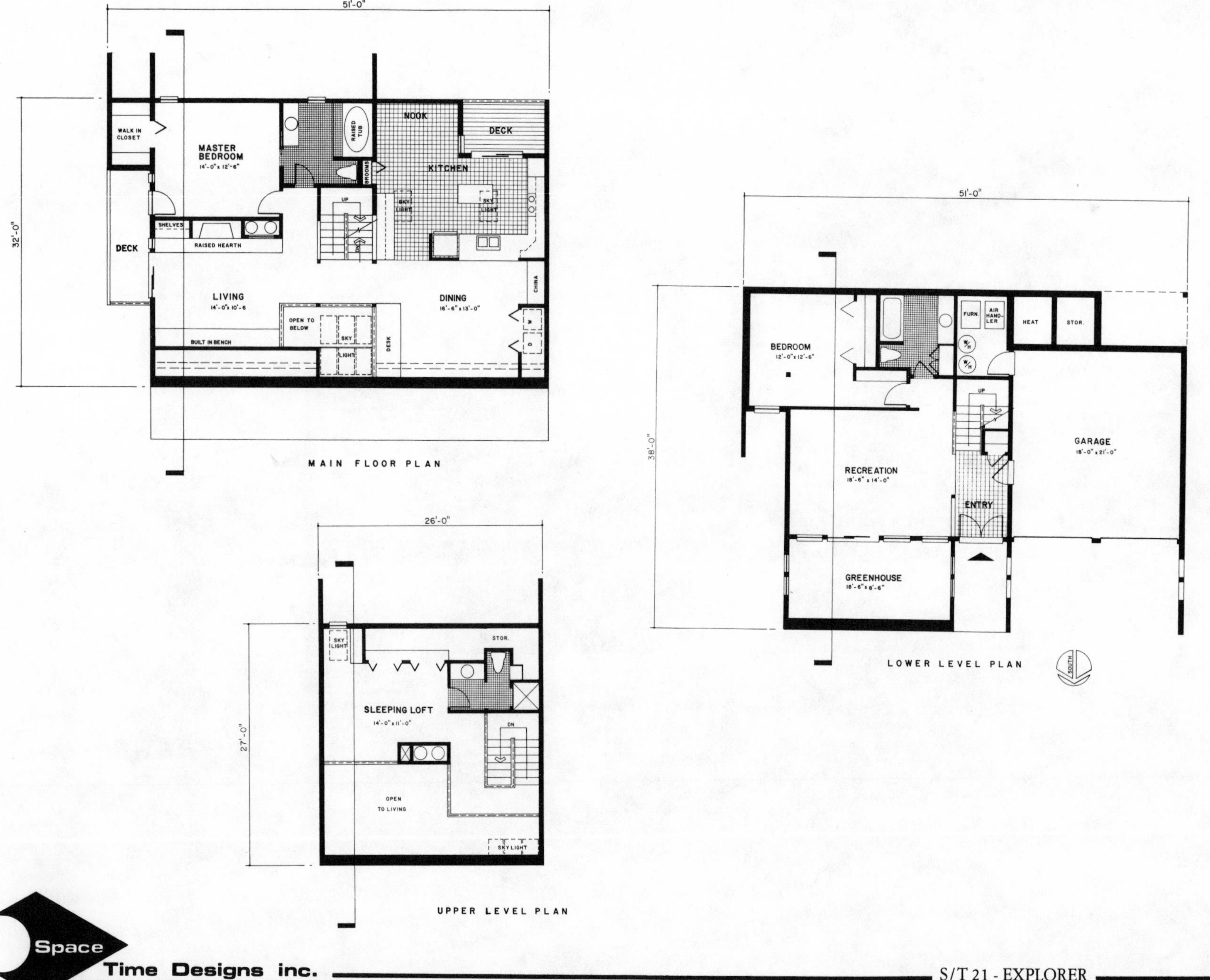

S/T 21 - EXPLORER

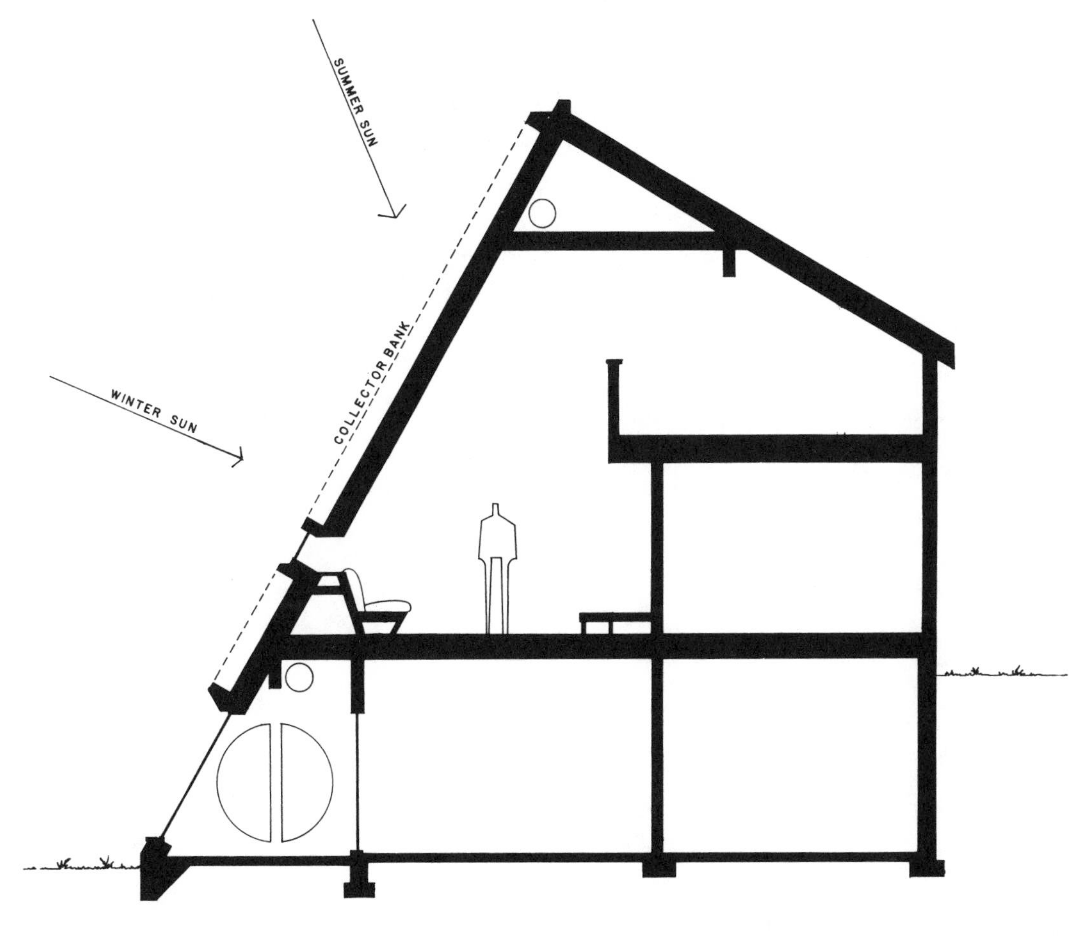

SECTION

DES MOINES
S/T-35

The handsome lines of this compact and practical 3-bedroom, 2½-bath home should look pleasing in any neighborhood. Additionally, the structure should be relatively simple and inexpensive to build. Set three feet below grade level, it gains an immediate energy advantage which is fully reinforced by the thermal envelope design with its 450 square foot solarium.

Functioning as an airlock entry, the solarium leads you into the open plan main floor with a large country kitchen and bright living and dining area. Utility areas buffer the house from the north and provide convenient access between house and attached garage.

Upstairs, all three bedrooms face onto the aesthetically delightful greenhouse space while the north-buffering baths and closets bring in their own illumination through skylights. Designed specifically to function well in relatively severe winter climates, this home is versatile enough to work well in a wide variety of geographic locations.

Main floor — 931 sq. ft.
Upper floor — 931 sq. ft.
Greenhouse volume — 9,120 cu. ft.
Total square footage — 1,862

Package Price — $550.00
3 bedrooms plus study
2½ bathrooms

Passive Features
Exterior vertical greenhouse glazing — 264 sq. ft.
Greenhouse 45° roof glazing — 312 sq. ft.
Optional breadbox solar water heater
Geo-thermal air envelope system

Active Features

Alternate water pre-heat system

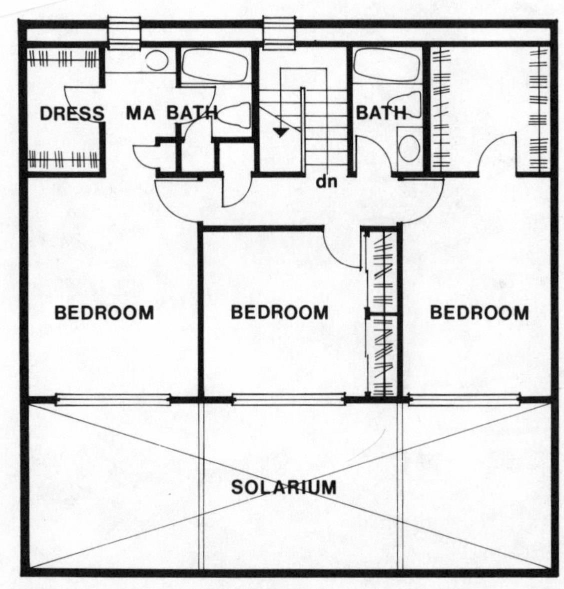

UPPER LEVEL PLAN

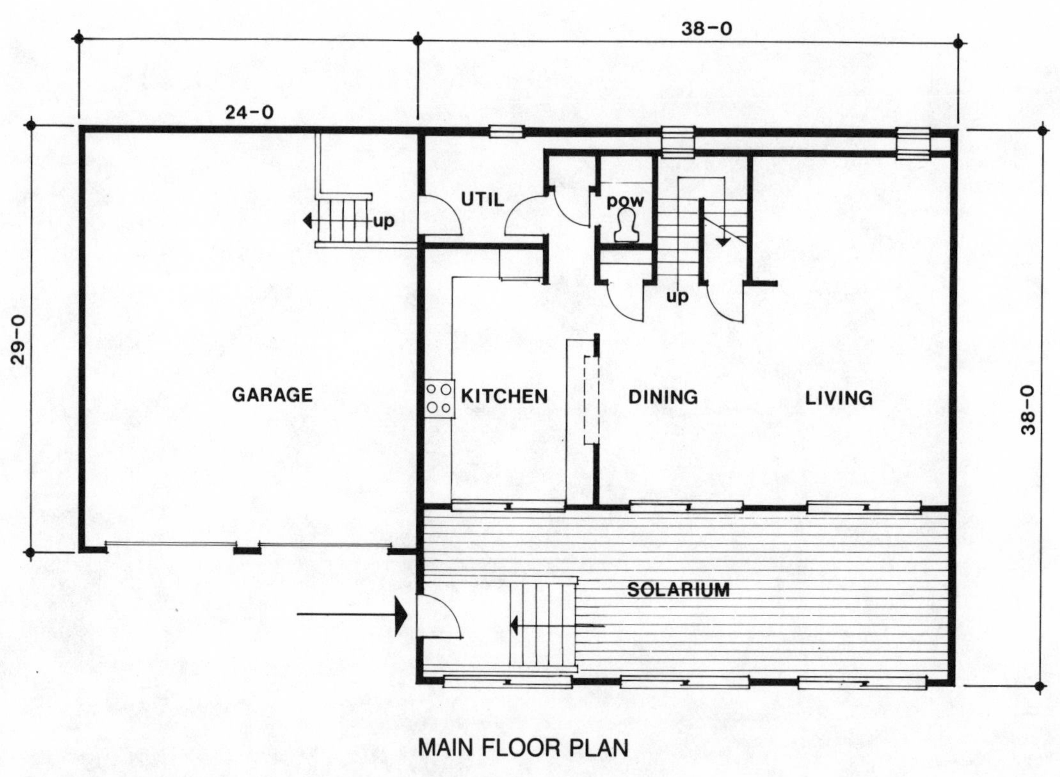

MAIN FLOOR PLAN

Space Time Designs inc.

S/T 35 - DES MOINES

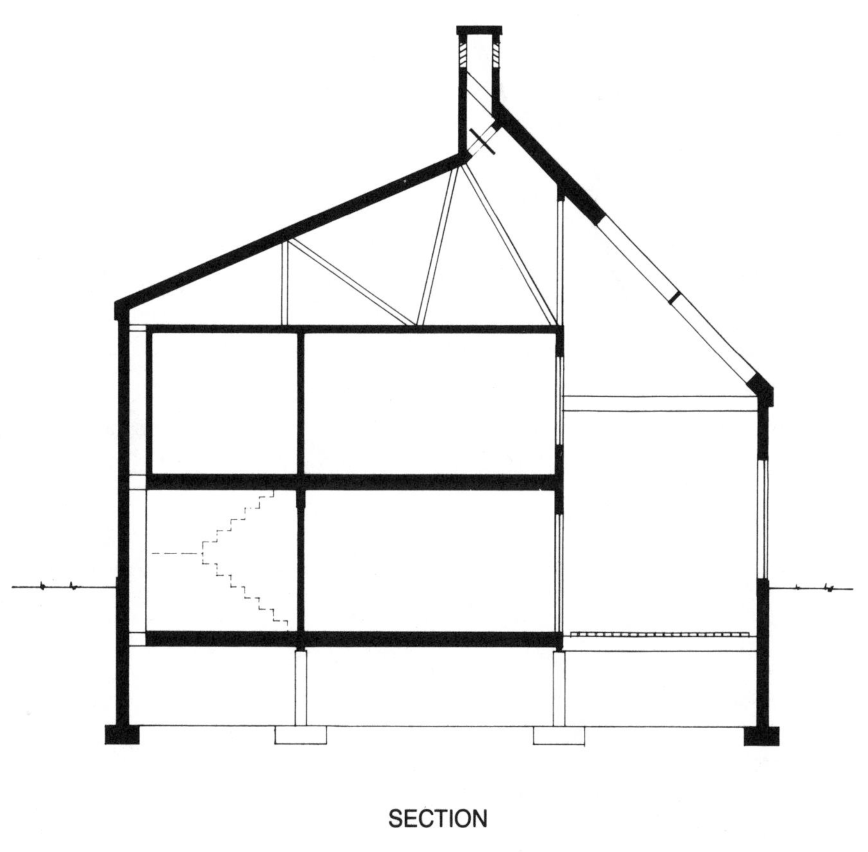

SECTION

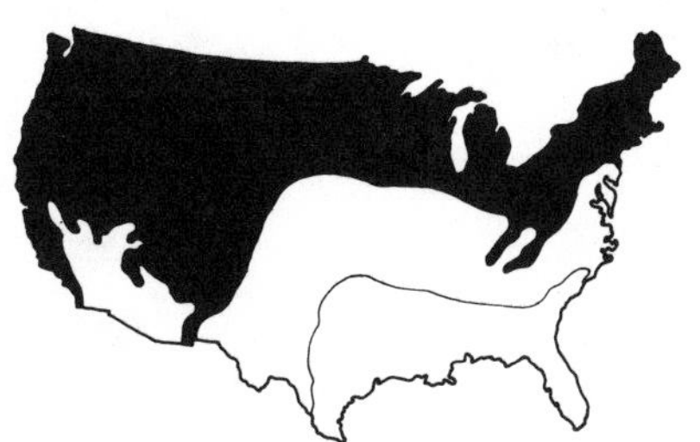

QUAIL'S NEST
S/T-36

Welcome the morning light as you awaken in your earth-protected master suite and arise to the naturally-retained heat of the main living area. Fix breakfast in the oversized kitchen and enjoy informal dining in the adjacent eating area as the interior mass wall continues to radiate its previous day's stored heat.

Designed for nest-like comfort and security, this home is ideal for any lot rising upward to the north with good sun exposure. Massive retaining walls reach out like arms to draw in the sun's warming rays to the core of the structure, which is approximately 68% earth sheltered. High, openable clerestory windows above the living room function both to heat a high mass wall and allow a pleasant summer cooling convection loop. Temperature-stabilizing masses are scattered uniformly throughout the house (master bedroom, bedroom #2, stairwell, hall, fireplace, and exterior structure as a whole), requiring little use of the suggested radiant slab heating back-up system except during extended periods of cloudy weather.

Ideal for a small family, this convenient one-story home combines energy efficiency, dynamic design, and simple living. Structural engineering for earth sheltered areas of this design must be calculated and supervised by a local licensed engineer, so the plan package includes reproduceable structural pages for his needs. No happy little quail family will ever have a snugger nest than this!

House square footage — 1,298
Garage square footage — 390

Package Price — $390.00
2 bedrooms plus den
2 bathrooms

Passive Features
Direct gain glazing — 226 sq. ft.
Potential direct mass storage — 180 cu. ft.
Structural mass storage — 1,450 cu. ft.
Approximately 68% earth sheltered

Active Features

Water pre-heat system
Potential exhaust fan in clerestory

M.L. ANDERSON '80

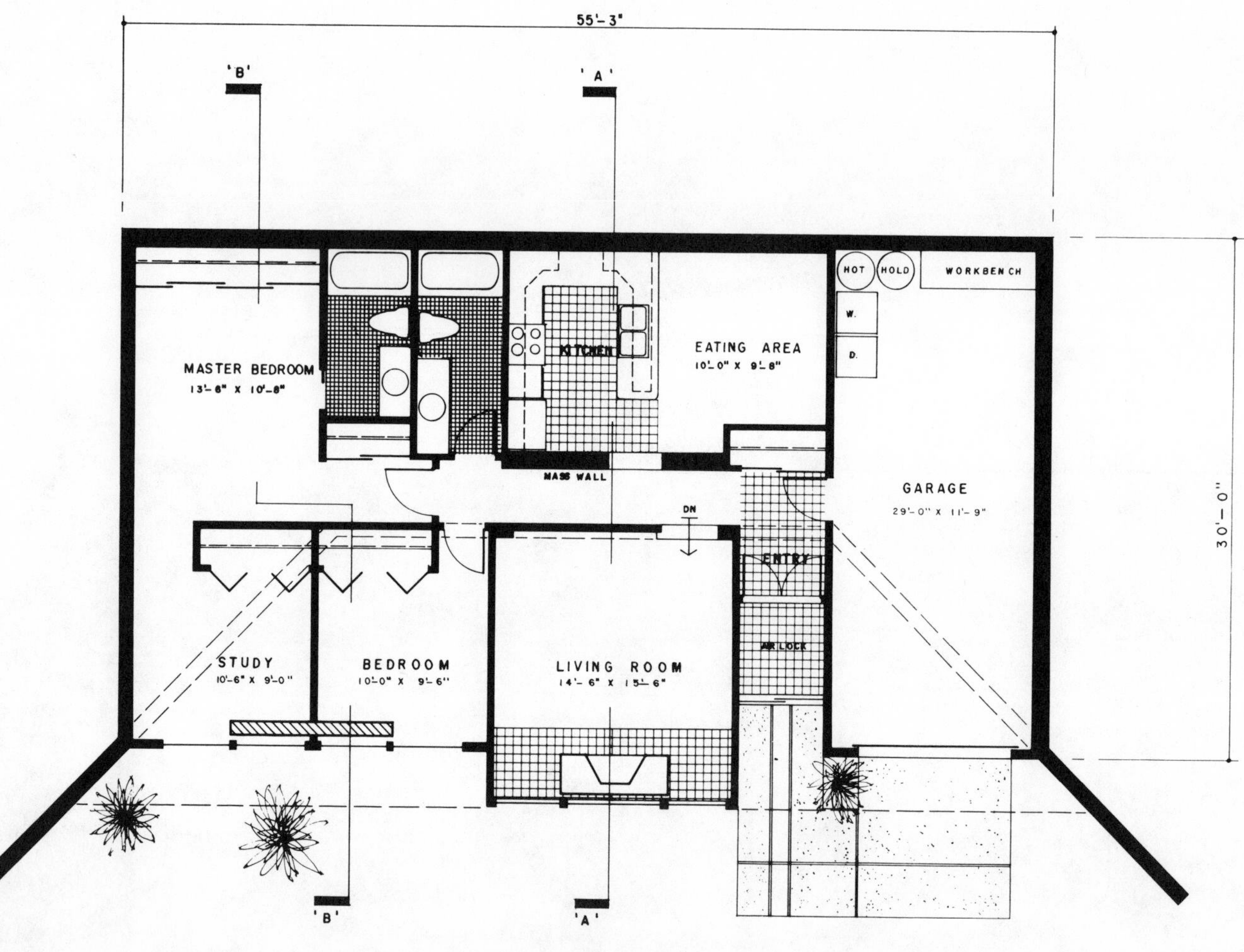

FLOOR PLAN

Space Time Designs inc.

S/T 36 - QUAIL'S NEST

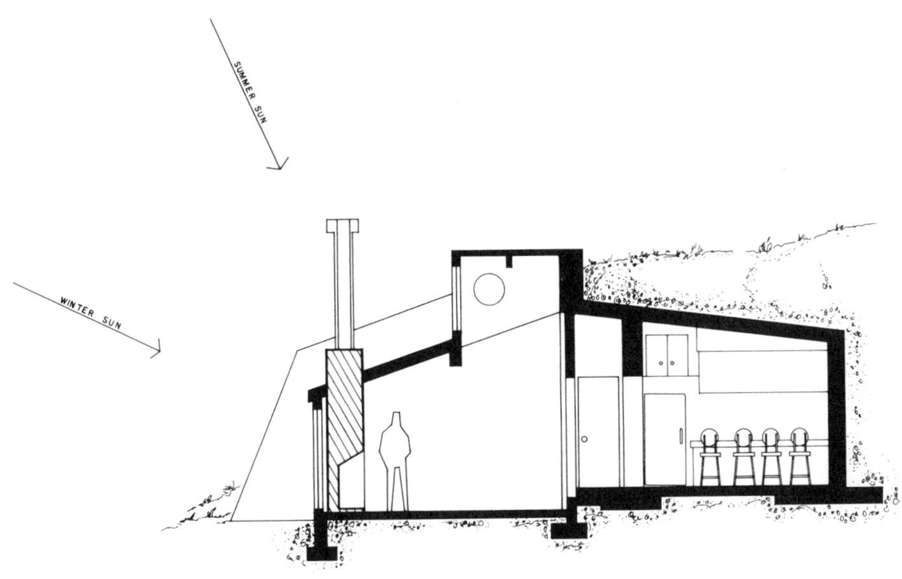

SECTION "A-A"

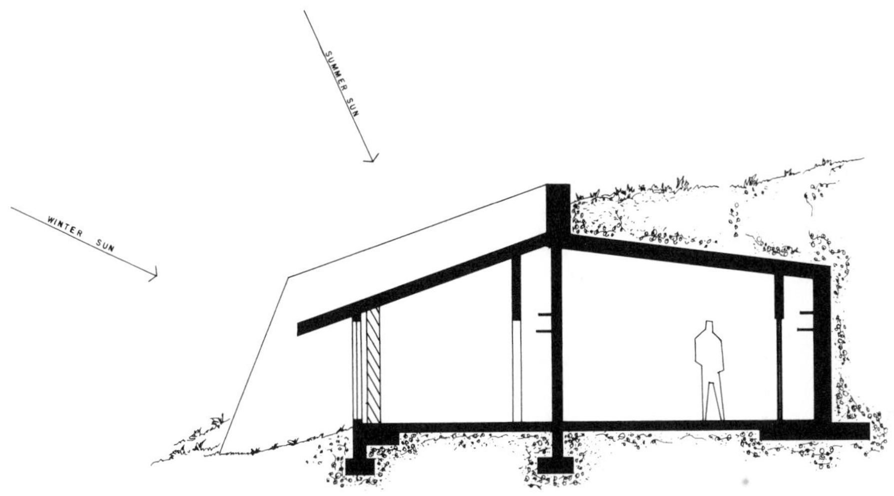

SECTION "B-B"

SUN CRUISER
S/T-24

Ideally suited for a north-sloping view site, this home can be a luxurious second home as shown with the roomy guest sleeping loft, or alternately, an ideal family home with a lofted master suite.* Proportioned to maximize energy conservation and solar energy collection simultaneously, the home promises a very high percentage of solar assistance. In addition, its purity of form assures simplicity of construction as well as beauty.

A feeling of openness permeates the upper floors; from the cozy raised den screened from the living room to the friendly open kitchen-dining area, the spaces flow easily from one to another. The stairway skylight pours light into the center of the house, and the two small downstairs bedrooms assure privacy for family members. A large outside storage room is ideal for storing extra firewood or lawn games and equipment.

If fewer collectors than we have shown are desired, the 60° tilt wall would be ideal for lighting and venting both the loft and main floor areas with openable skylights. Basement active storage ensures that stored heat will remain within the structure. A 20'x24' garage is included, but not shown in the plan view.

*Be sure to specify which version is preferred when ordering plans. Use number S/T-24A if master suite is desired.

Main floor — 822 sq. ft.
Lower floor — 896 sq. ft.
Upper floor — 448 sq. ft.
Total square footage — 2,166

Package price — $425.00 with garage
3 bedrooms plus den
3 bathrooms

Passive Features
Summer ventilation

Active Features
Potential collector area — 620 sq. ft.
Potential storage volume — 440 cu. ft.
Collector tilt — 60°
Suggested water pre-heat system is an air-to-water heat exchanger

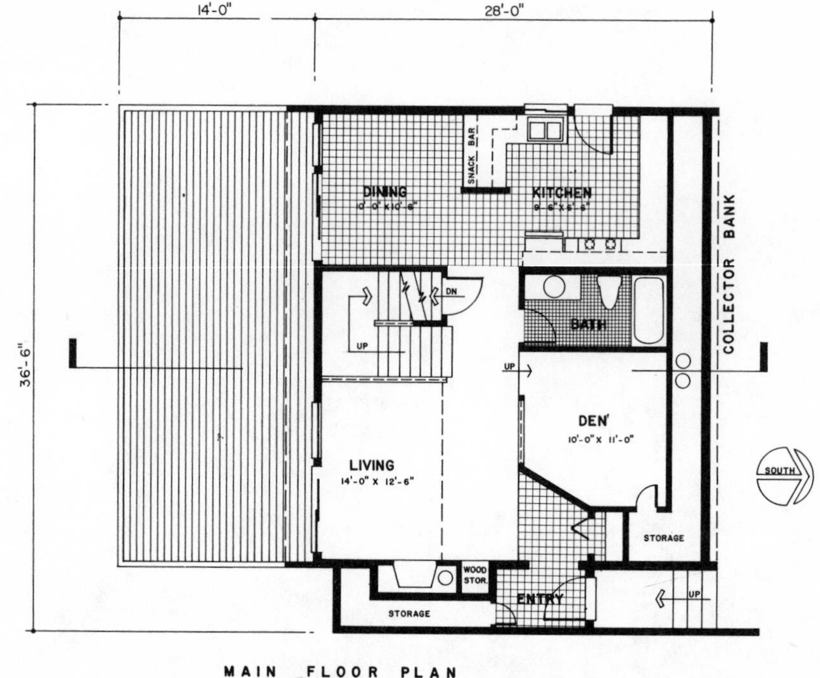

MAIN FLOOR PLAN

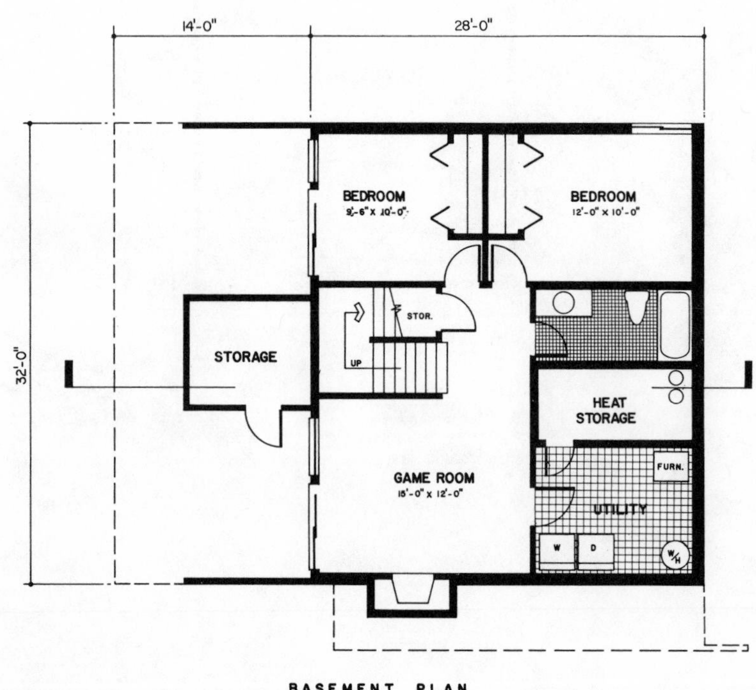

BASEMENT PLAN

Space Time Designs inc.

S/T 24 - SUN CRUISER

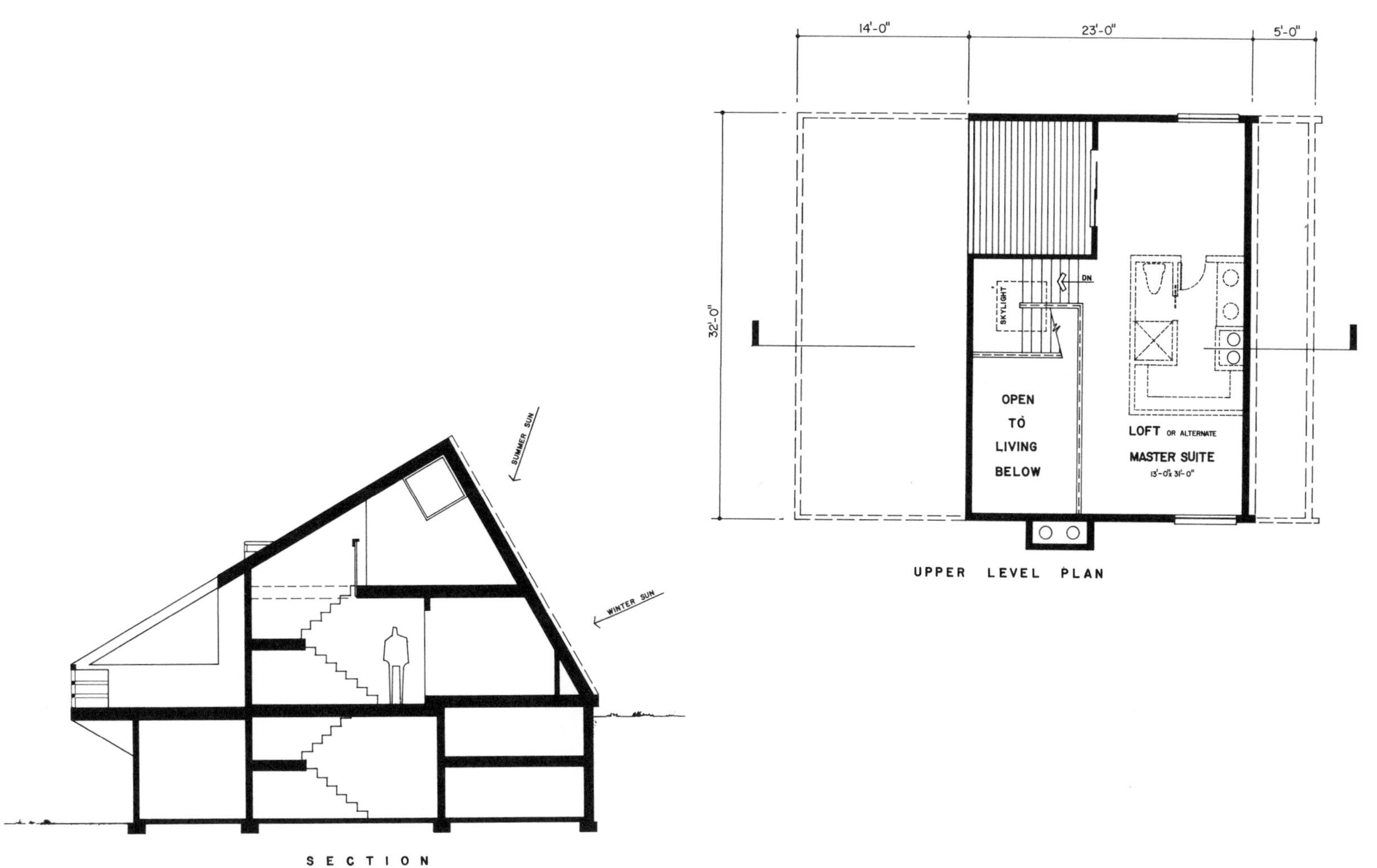

14'-0"
23'-0"
5'-0"
32'-0"
SKYLIGHT
DN
OPEN
TO
LIVING
BELOW
LOFT OR ALTERNATE
MASTER SUITE
13'-0" x 31'-0"
UPPER LEVEL PLAN
SUMMER SUN
WINTER SUN
SECTION

CLASSIC FUNK
S/T-25

Astonishingly funky and eclectic throughout, this house exists in a dimension beyond time.

A dynamic curving skylight crowns the sculptured entrance porch, inviting you inside an open sun-heated activity area adjacent to the master suite and bath.

An open dining atrium divides a roomy kitchen and sunken livingroom, each with private balconies, to complete this total viewfront living space that pleasantly opens onto the greenhouses.

A radiating wood shake sunburst surrounds the energy efficient fireplace, adding to the spatial excitement of this interior wonderland.

Downstairs, two bedrooms - one opening to a private patio and the other to one of the two greenhouses - share a large family bath, and have easy access to the utility room.

A large recreation room opens to the second greenhouse which, along with a rustic fireplace, combines to warm this space during the winter season.

Designed for people with a laid-back lifestyle, with images of the classics presented in a homegrown spirit, this house promises basic funky family living.

Main floor — 1,161 sq. ft.
Upper floor — 1,065 sq. ft.
Greenhouse volume — 3,260 cu. ft.
Total square footage — 2,226

Package price — $610.00
3 bedrooms
2½ bathrooms

Passive Features
Direct gain glazing — 84 sq. ft.
Greenhouse glazing — 336 sq. ft.
Potential mass storage — 1,000 cu. ft.

Active Features
Water pre-heat system

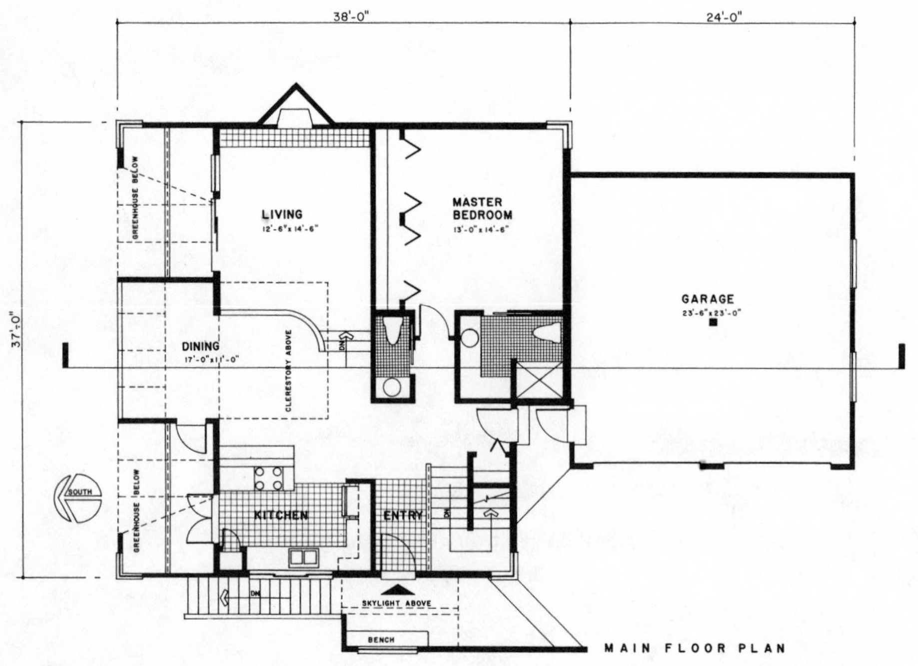

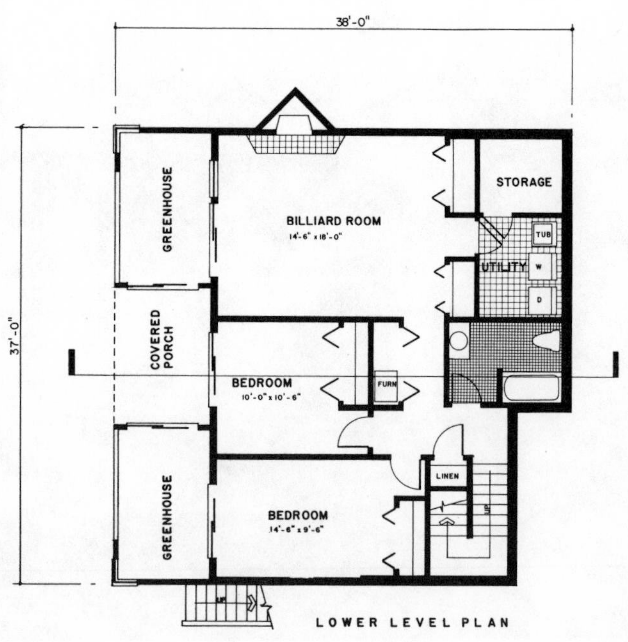

Space
Time Designs inc.

S/T 25'- CLASSIC FUNK

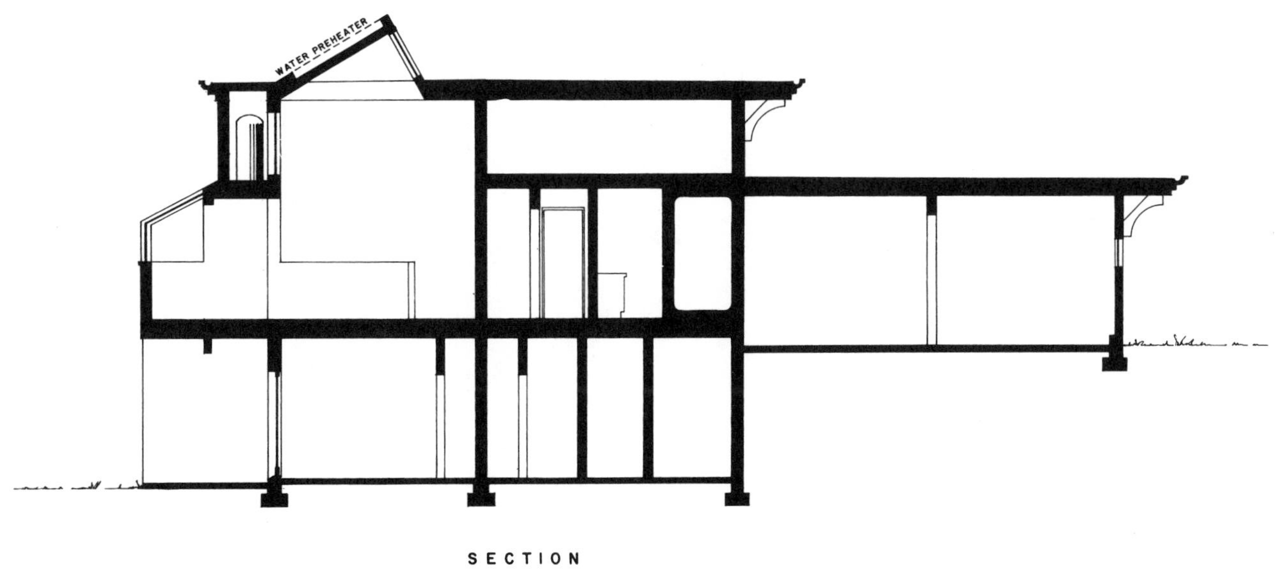

SECTION

THE VAGABOND'S CABIN
S/T-26

Designed for economical building on a woodland or beach front site, this informal retreat with its ample collector areas and heat-efficient stove will rarely need its back-up furnace.

The reflective surface on the flat roof area will reflect hot summer rays back to the sky and divert additional winter sun into the collectors. This would nearly double the efficiency of the collectors if a 90% reflective coating is applied over an asphalt multi-ply roof, providing you with an effective collector area of over 250 square feet.

Imagine the possibilities: cantilever the deck out over a river or above rolling sand dunes. Use naturally heated water for washing dishes and people; retreat to a cool, private master bedroom; gather kids and adults around a crackling fire; sleep a crowd in the high-ceilinged bunkroom.

Main floor — 930 sq. ft.
Upper floor — 392 sq. ft.
Total square footage — 1,322

Package price — $310.00
1 bedroom + loft
2 bathrooms

Passive Features
Direct gain glazing — 105 sq. ft.
Potential mass storage — 200 cu. ft.

Active Features
Potential collector area — 135 sq. ft.
Potential storage volume — 375 cu. ft.
Collector tilt — 90o
Water pre-heat system

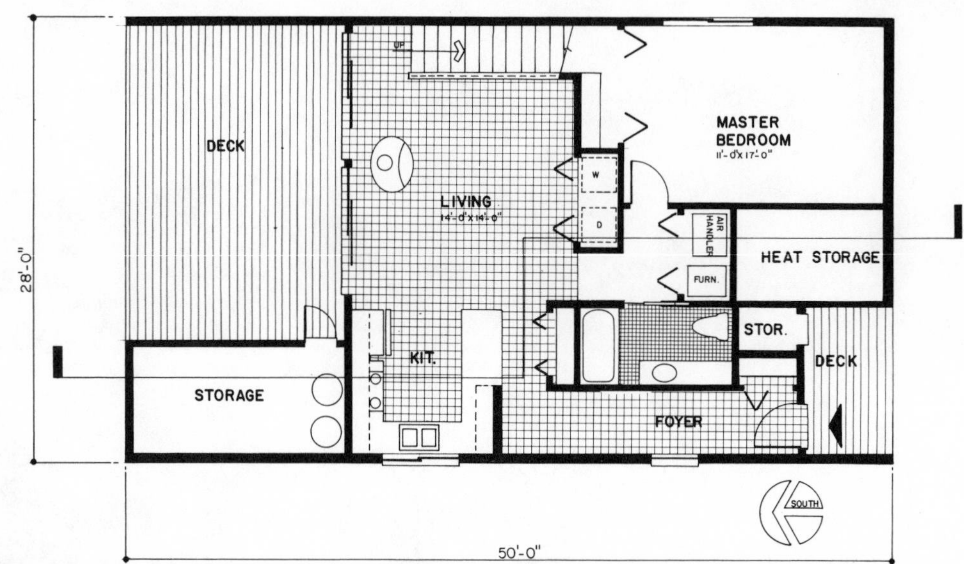

MAIN FLOOR PLAN

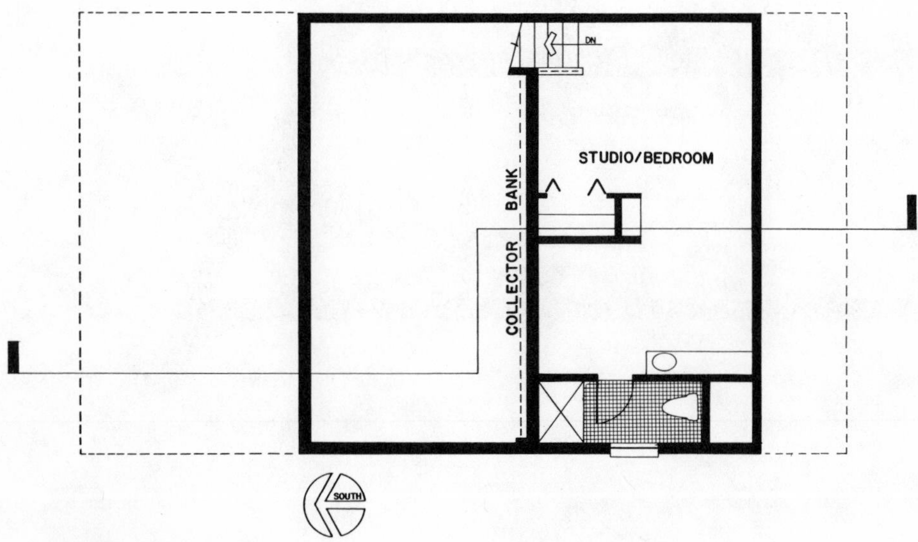

UPPER LEVEL PLAN

Space
Time Designs inc.

S/T 26 - THE VAGABOND'S CABIN

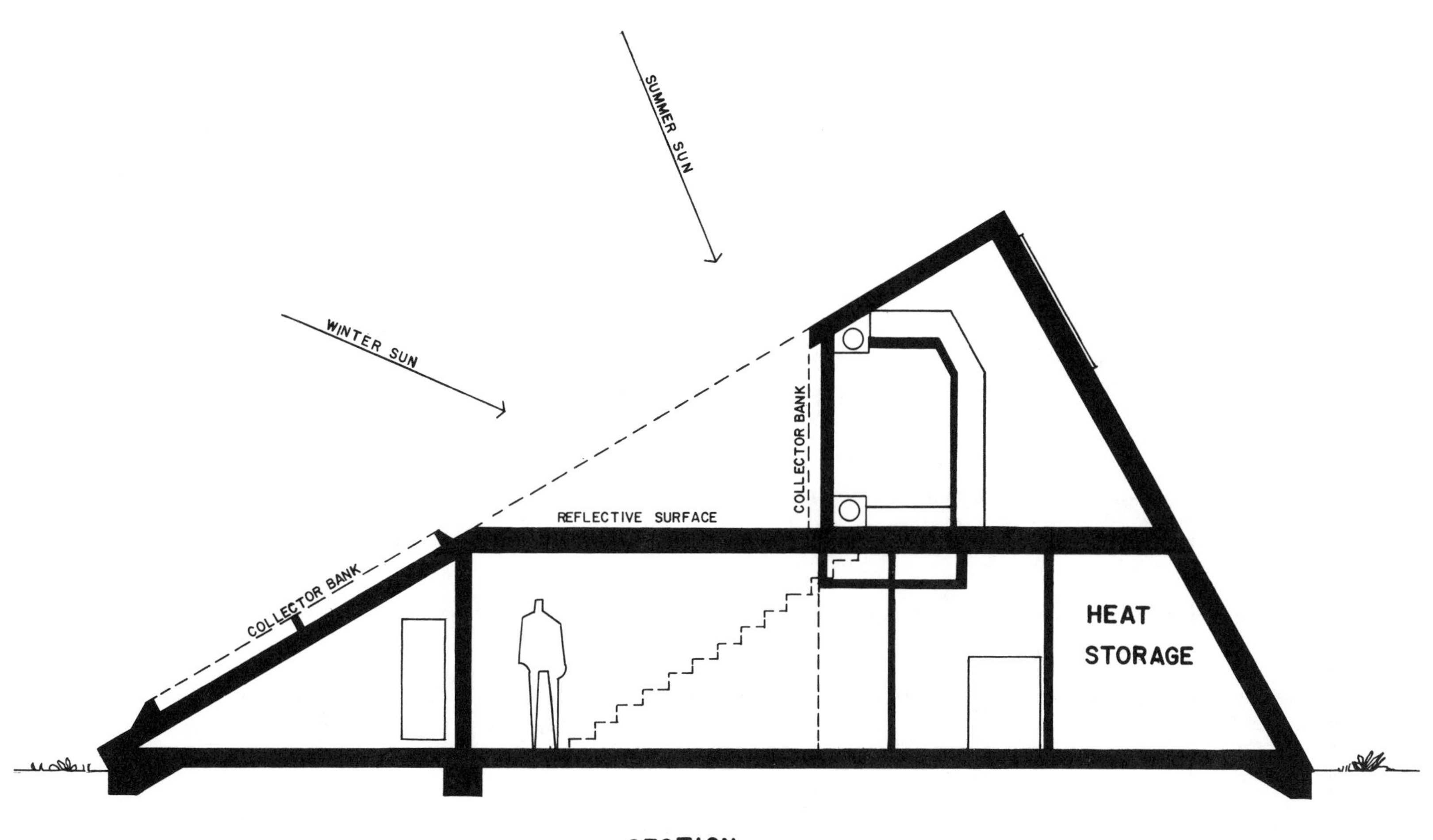
SUMMER SUN
WINTER SUN
COLLECTOR BANK
REFLECTIVE SURFACE
COLLECTOR BANK
HEAT
STORAGE

SECTION

SPACEDOME II
S/T-37

Ideally suited for a downward south slope, this 90% earth sheltered hemisphere is an open-faced concrete shell structure with exciting innerspaces. Featuring a level entry into the upper story living and master bedroom areas, the entire floor plan is designed for convenience and effective zoning. In the sculptural master suite, imagine the mirrored bipass closet doors reflecting a tunnelled cedar ring skylight over your king-sized waterbed, and leading you into a sensuous bath with a luxurious raised tile tub. A large dining area open to the stairwell is conveniently located for entertaining between the kitchen and living room. Curved cedar and plexiglass windows bent around a free-standing wood-burning stove invite woods, sun and view inside.

Down a spacious open stairwell are a centrally-located bath, three more interestingly shaped bedrooms and an ample sewing room which draws light from an open well under the entry bridge. Leading outside from the open playroom are retaining walls outstretched to gather solar gain and store it in the mass storage floor slab for nighttime reradiation. All downstairs bedrooms have outside entrances.

Although code requires that you have this home locally engineered and supervised, we believe that it will provide you and your family convenience, comfort, and enlightening and playful spatial forms. Pre-prepared structural and forming engineering sheets will be provided with this plan package, as are the plans for Cardome 30.

Main floor — 1,086 sq. ft.
Lower floor — 1,358 sq. ft.
Total square footage — 2,444

Package Price — $780.00
4 bedrooms plus sewing area and playroom
2 bathrooms

Passive Features
South-facing glass — 138 sq. ft.
Approximately 90% earth sheltered
Structural mass — 880 cu. ft. concrete shell

Active Features
Water pre-heat system

PETERSON '80

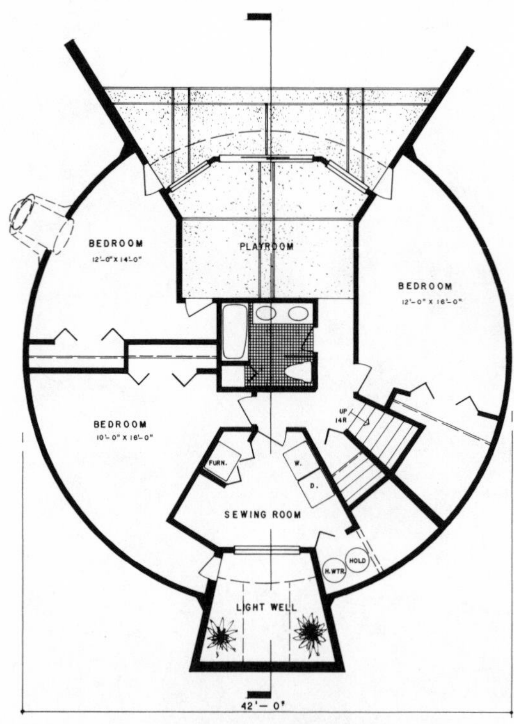

LOWER FLOOR PLAN

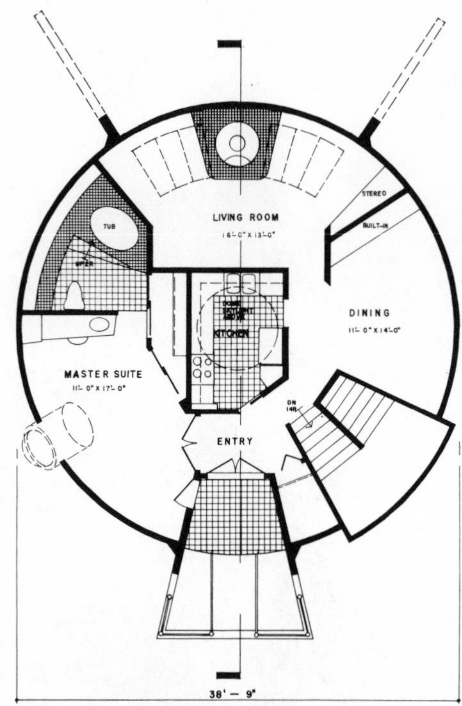

UPPER FLOOR PLAN

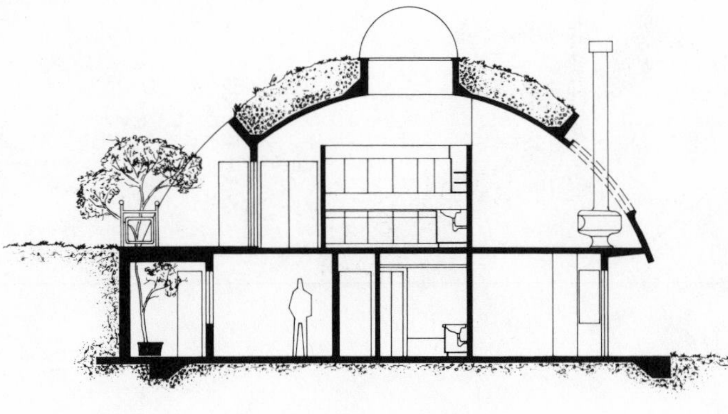

SECTION

For garage plan and section see pages 70 and 71.

Space Time Designs inc. S/T 37 - SPACEDOME II

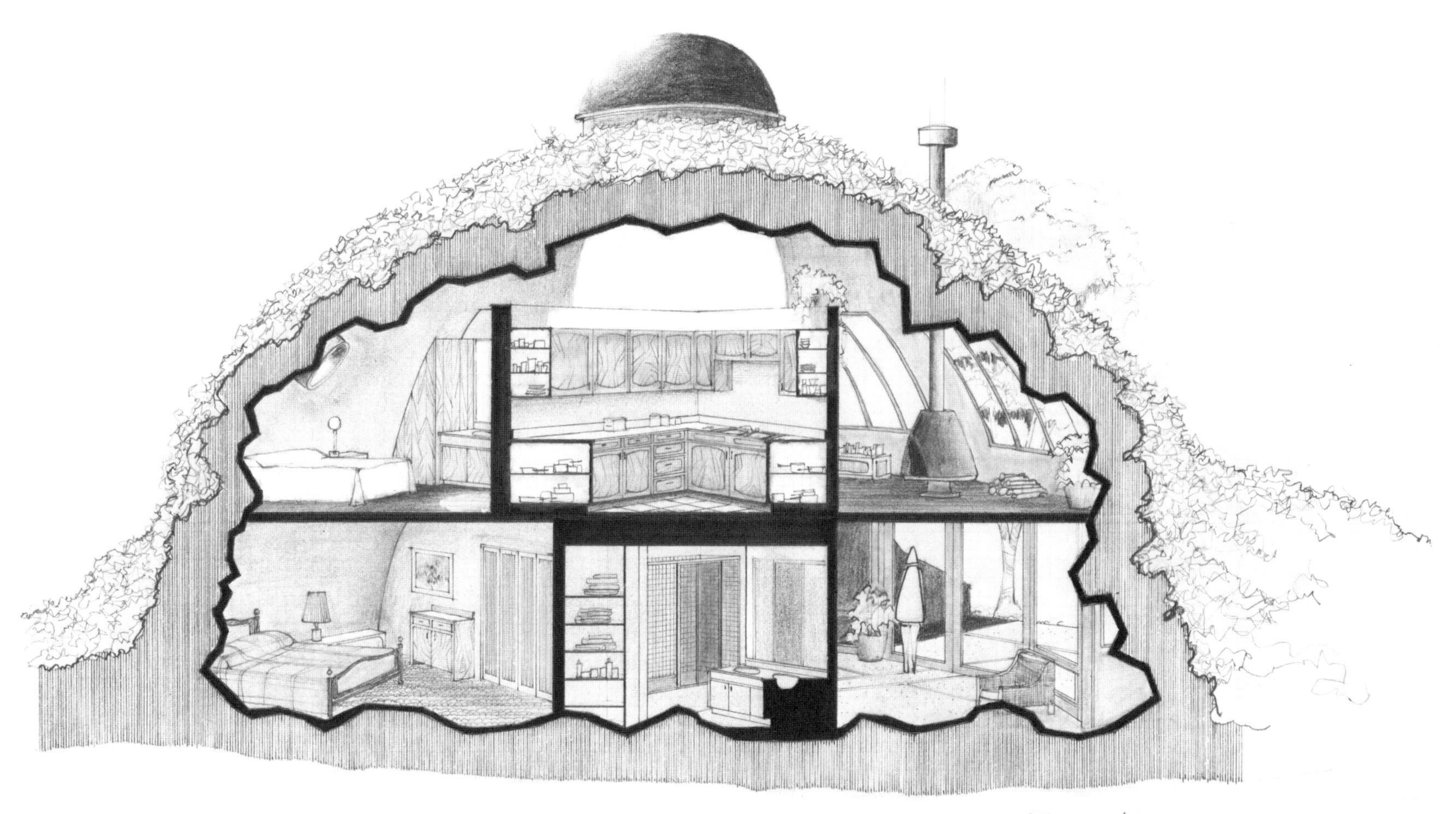
PETERSON '80

AGE OF LUXURY
S/T-28

A centralized two-story entry glowing with stained glass leads you into a house at once gloriously self-indulgent and extremely practical and efficient. There's a huge master suite, and a convenient kitchen off the two-level greenhouse which has space enough at the west end for a sunny jungle breakfast. A formal dining room also adjoins the greenhouse and opens to the soaring spaces of the living room to the north. Downstairs is the play area for family and friends. Take a sauna or a soak in the hot tub, play pool in the enormous game room, sip a glass of wine, and enjoy your lush botanical garden from all these spaces.

The greenhouse can accommodate any of a variety of mass storage receptacles which will radiate stored sunshine back into the house at night. Additional engineering may enable you to direct fan-driven warm air to rock storage in the crawl space, and thus provide an added BTU source for the central forced air furnace. With some sort of insulated night-time cover over the greenhouse glazing, this design could fill a large percentage of your total heating requirements. A 20' x 22' detached garage (similar to S/T-16) permits a wide variety of siting choices. And you may even be able to charge admission to your botanical gardens!

Main floor — 1,272 sq. ft.
Lower floor — 1,363 sq. ft.
Greenhouse volume — 2,340 cu. ft.
Total square footage — 2,635

Package price — $680.00
3 bedrooms
2½ bathrooms

Passive Features
Greenhouse glazing — 600 sq. ft.
Potential mass storage — 800 cu. ft.

Active Features
Suggest excess heat be fan-driven to crawl space rock bin
Suggested water pre-heating by locating single glazed collectors in greenhouse next to kitchen

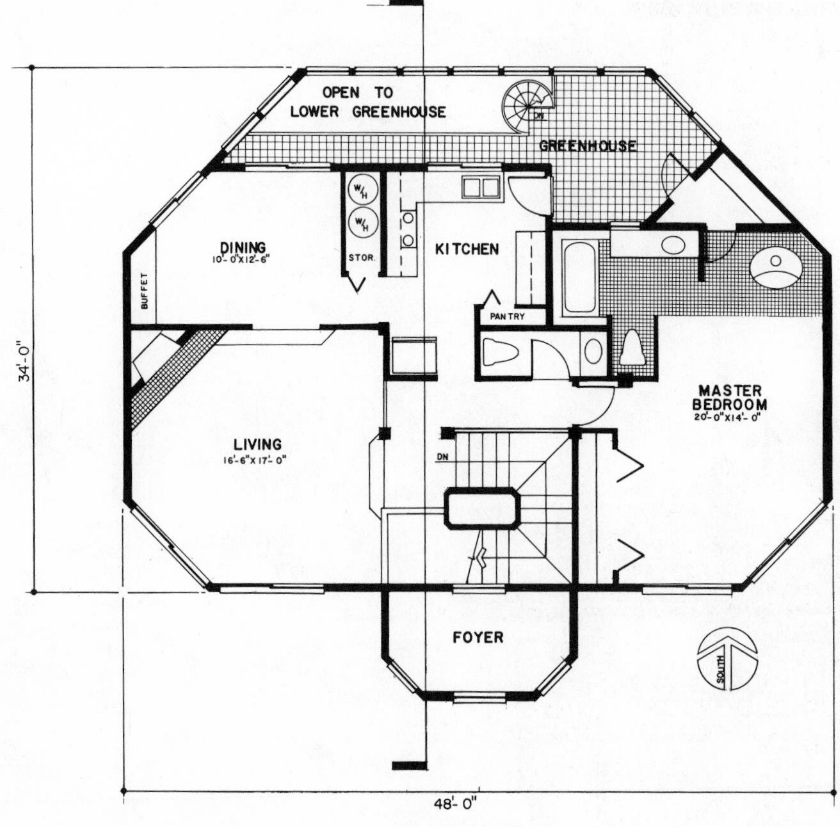

MAIN FLOOR PLAN

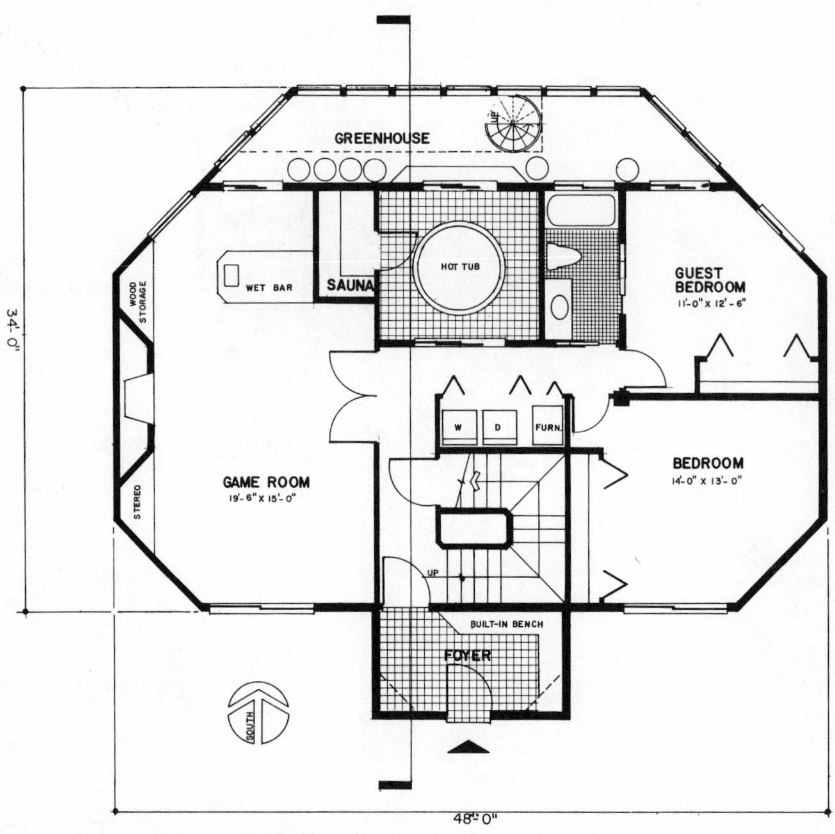

LOWER FLOOR PLAN

Space Time Designs inc.

S/T 28 - AGE OF LUXURY

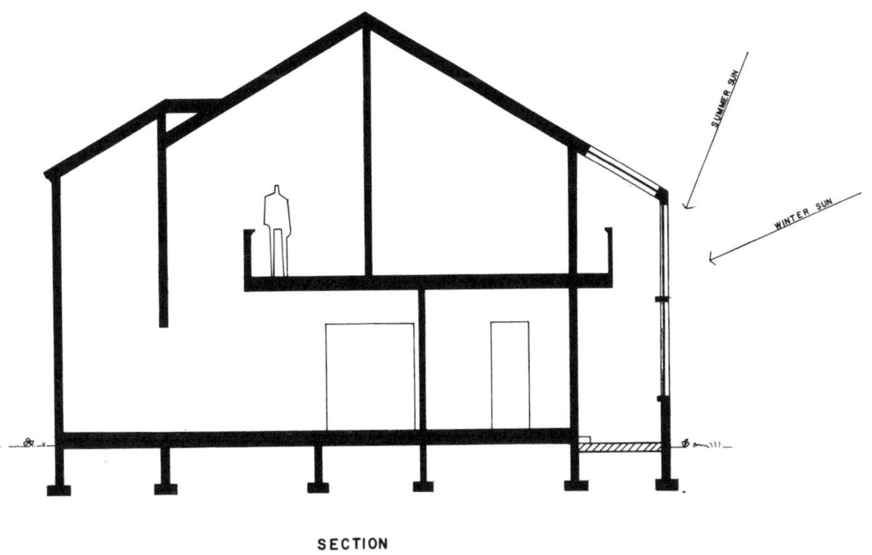

SECTION

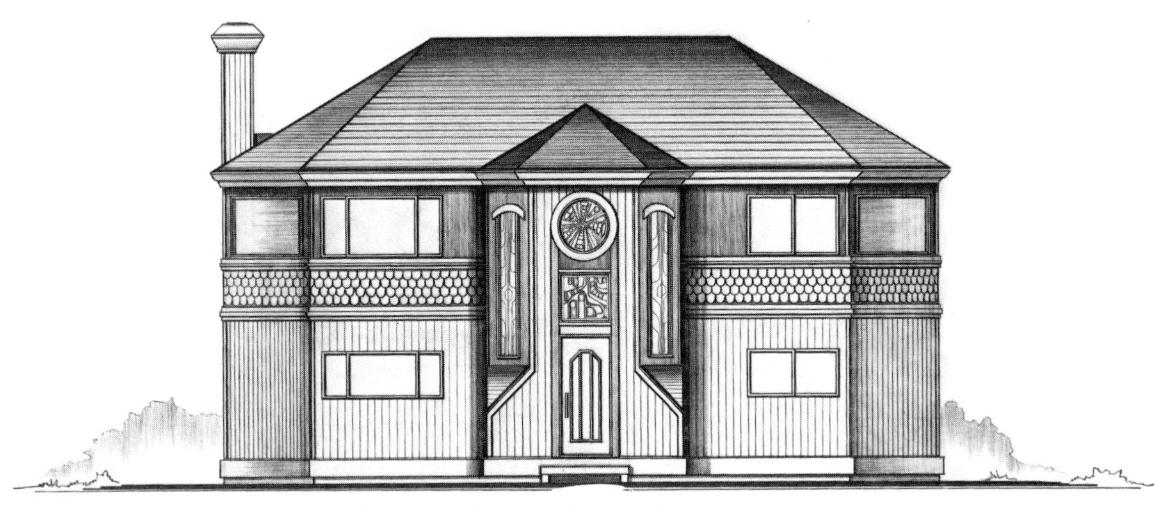

NORTH ELEVATION

APOLLO
S/T-29

Enriched with stained glass and sculptured wood shakes, this house is much less complicated than it might at first appear. Organic, homegrown and funky, the design promises fun living as well as effective solar heat assistance.

Having more than 700 square feet of solar aperture and well protected to the north by the garage, storage area, and sloping roof, this house will retain a high amount of solar energy. East/west windows should permit easy summer temperature control by exhausting warm air through the north-facing clerestory. The centrally located living room fireplace is positioned so that it could accomodate some additional storage capacity if planned carefully. A heat exchanger in the fireplace or in the path of solar air collection is the suggested water pre-heat method.

Patchwork wood paneling is detailed to compliment the rustic wood screens which informally set off the living room. The utility area and entry to the garage are conveniently located in the compact kitchen which features a built-in island for cooking and snacks. Heat assistance as well as greenhouse beauty is provided through sliding patio doors off the living room, dining room and master suite.

The rustic nature of this home promises enhanced living as well as a solidly functional floor plan. With clustered bedrooms, it is ideally suited for the young family.

Main floor — 1,030 sq. ft.
Upper floor — 997 sq. ft.
Greenhouse volume — 1,280 cu. ft.
Total square footage — 2,027

Package price — $620.00
3 bedrooms plus den
2½ bathrooms

Passive Features
Greenhouse glazing — 224 sq. ft.
Potential mass storage — 400 cu. ft.

Active Features
Potential collector area — 504 sq. ft.
Potential storage volume — 416 cu. ft.
Collector tilt — 60°
Suggested water pre-heat system is an air to water heat exchanger

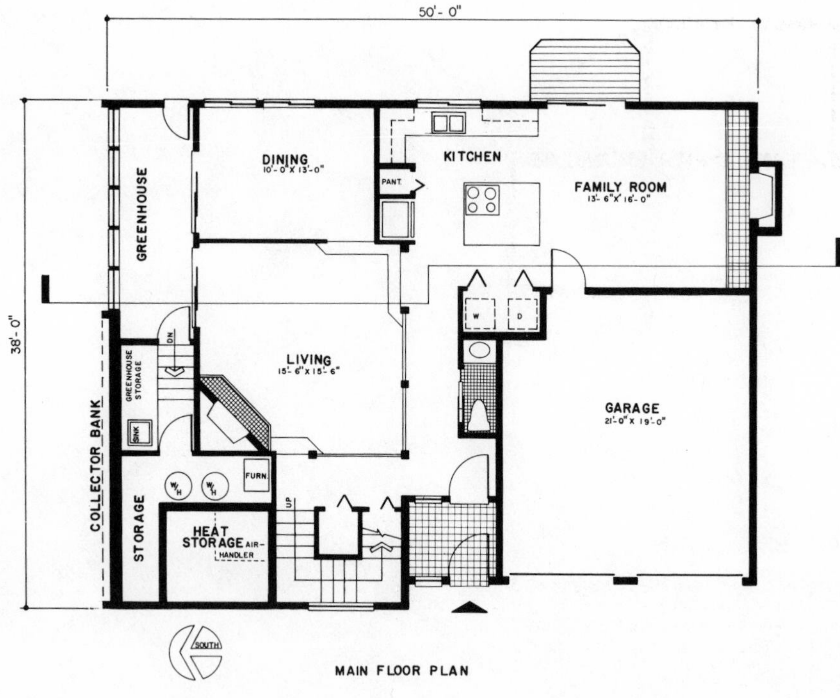

MAIN FLOOR PLAN

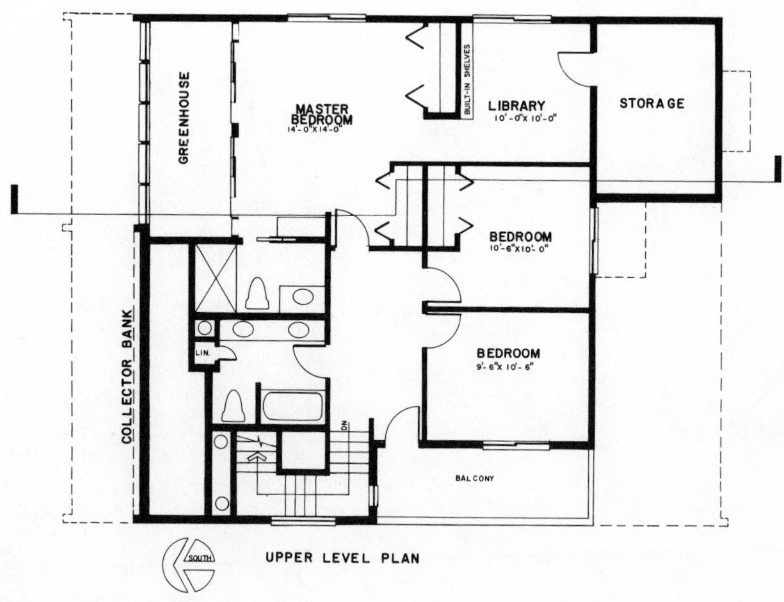

UPPER LEVEL PLAN

Space Time Designs inc.

S/T 29 - APOLLO

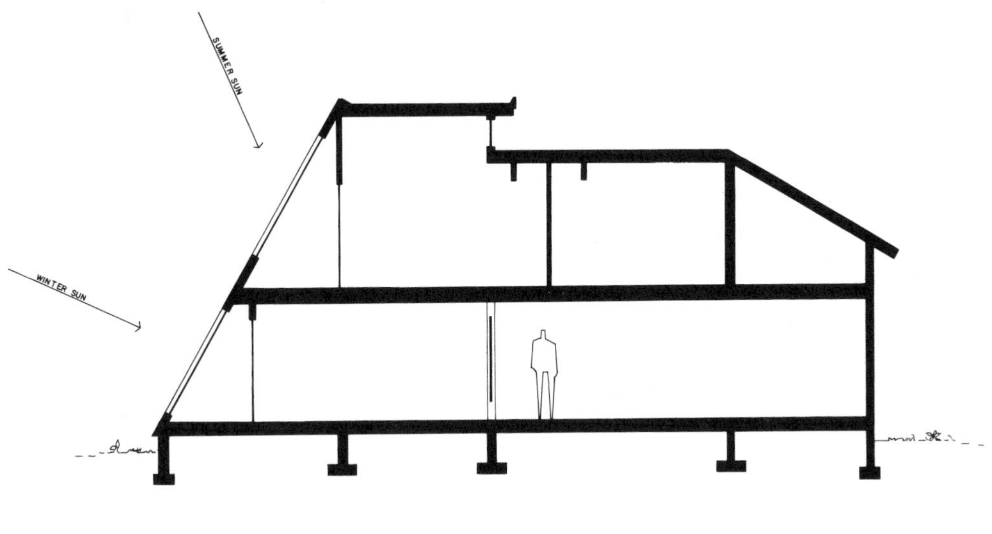

SECTION

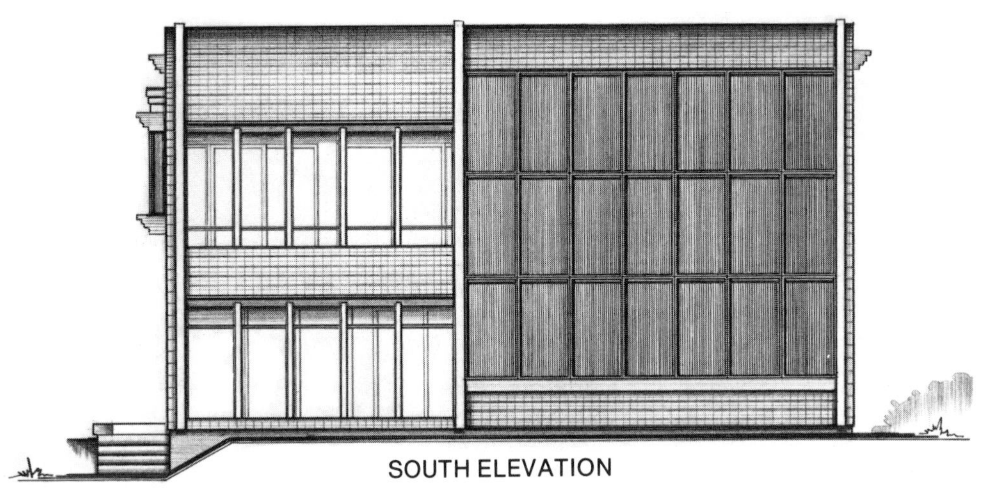

SOUTH ELEVATION

XP 230-7
S/T-30

This giant crawling sungrabber flowing up and over the rear north slope reveals little of the warmth and light which pours into the interior and turns the house into a series of lush hanging gardens. Conceived by Dave Fordon, the dynamic complexity of the house is centered by the massive direct gain atrium over the abstract sculpture of the fireplace conversation area. Radiating from and open to this stunning focal point are the quiet living room, dining room overlooking the tilt-glazed mini-garden, studio, and convenient kitchen. Screened from the north by a utility and storage wall, the entire home is illuminated by the expansive skylighting set under adjustable louvers for year-around sun control.

Winding up the open staircase past the home's second mini-garden, you enter the lofted second level overlooking the atrium. A spacious master suite above the mini-garden shares a large deck and elegant bath with the second bedroom.

A dream home for an adventuresome botanist or an interior decorator, this design is ideally suited to an upward-sloping, south-facing hillside in a quasi-wooded setting. Carefully placed round windows - one in the stairwell and the other split between master bedroom and dining room - will frame east and west views unique to your site.

Main floor — 1,535 sq. ft.
Upper floor — 873 sq. ft.
Total square footage — 2,408

Package price — $720.00
2 bedrooms plus study
1½ bathrooms

Passive Features
Direct gain glazing — 576 sq. ft.
Potential mass storage — 800 cu. ft.

Active Features
Potential collector area — 324 sq. ft.
Potential storage volume — 400 cu. ft.
Collector tilt — 90°
Suggested water pre-heat system is an air to water heat exchanger

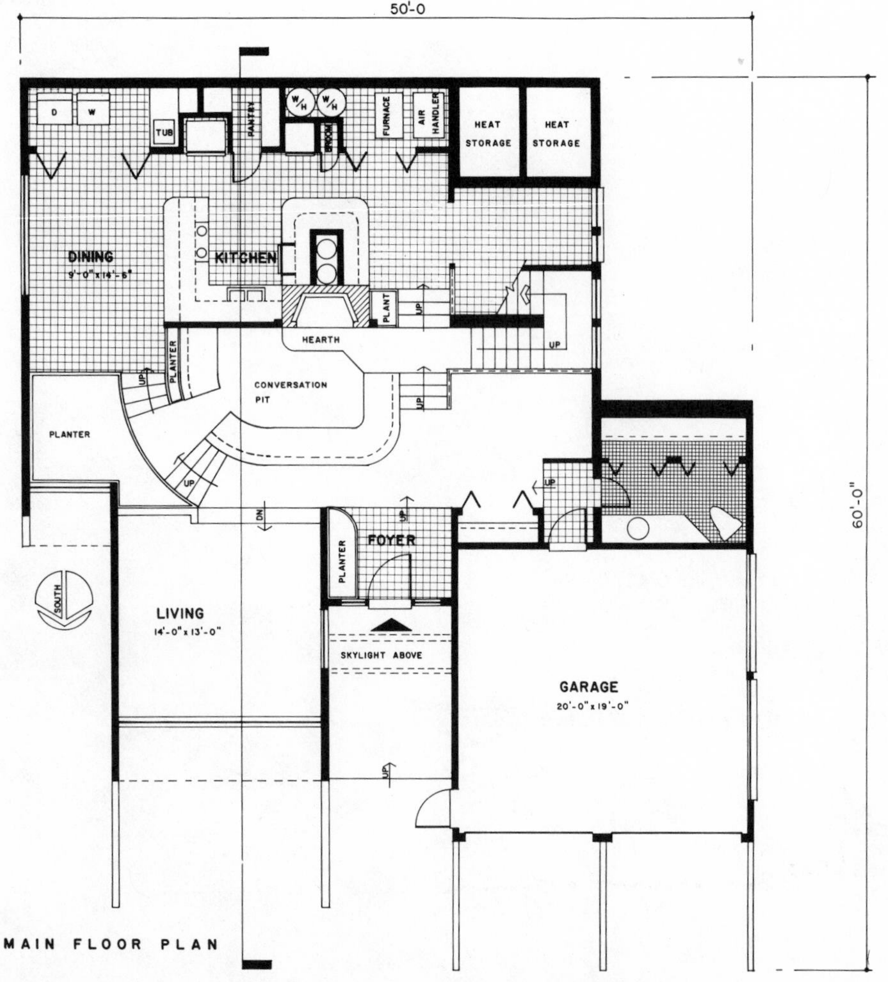

MAIN FLOOR PLAN

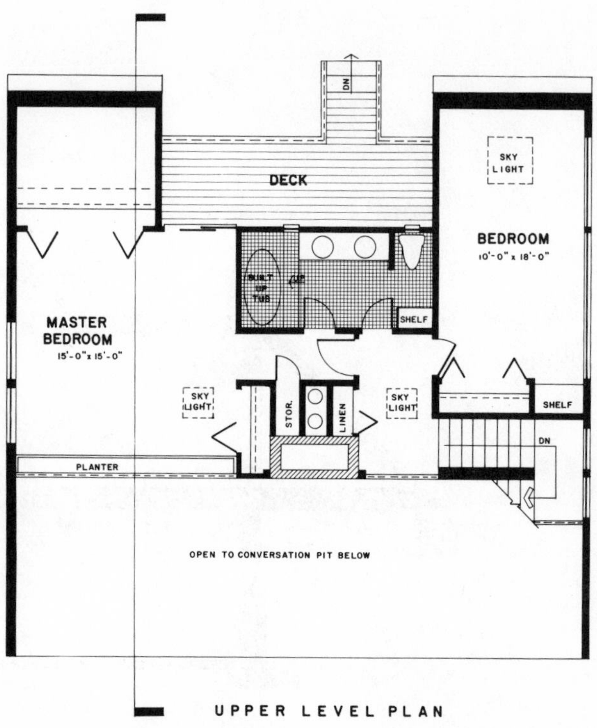

UPPER LEVEL PLAN

Space Time Designs inc. S/T 30 - XP 230-7

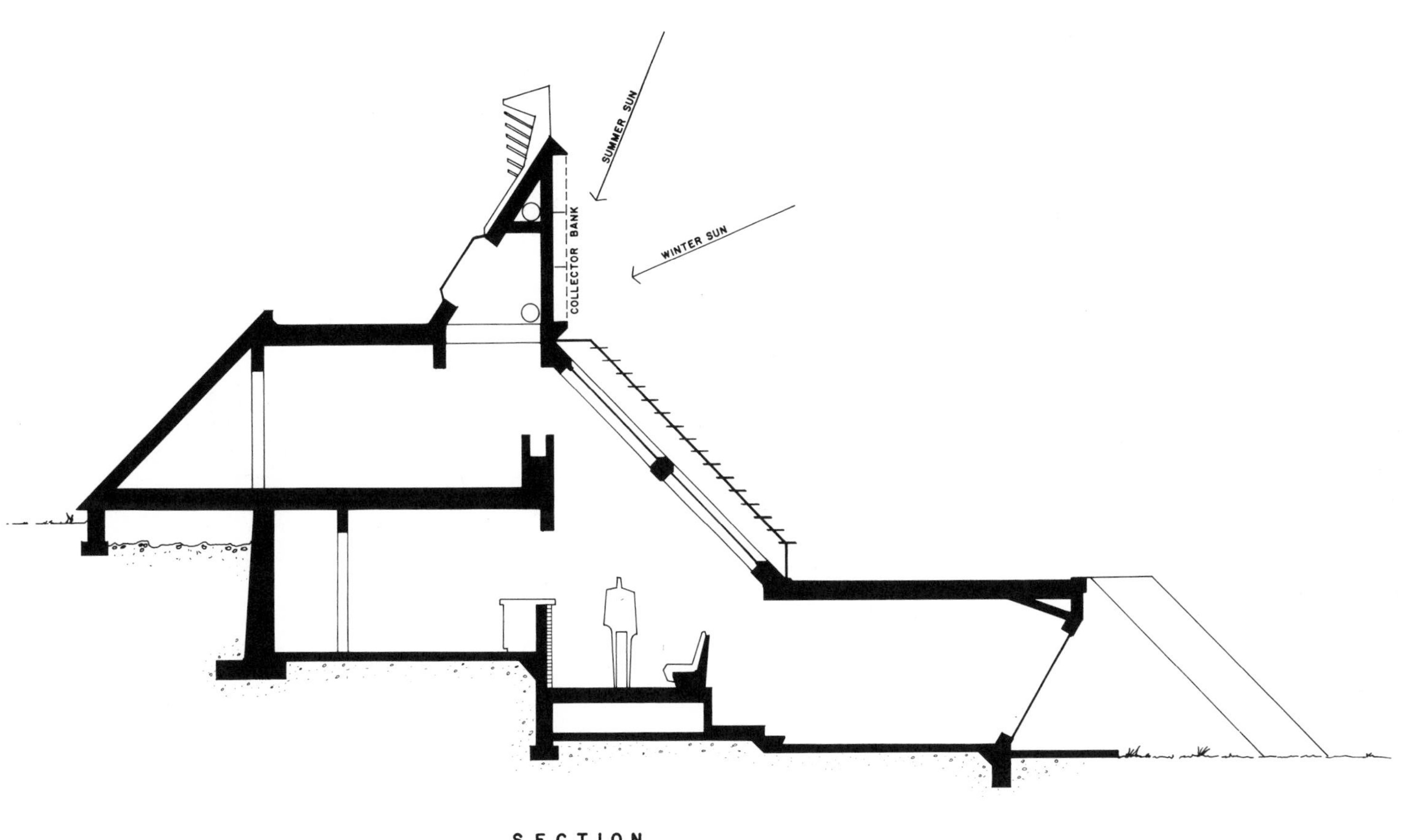

SECTION

SPACE/TIME DESIGNS INC. ORDER CHART

PLAN NO.	PLAN NAME	SF AREA MAIN	SF AREA LOWER	SF AREA UPPER	TOTAL SF	SF ACTIVE COLLECTOR AREA	SF PASSIVE APERTURE	TOTAL SOLAR APERTURE	PACKAGE PRICE	ADDITIONAL NOTES
S/T 1	THE OVEN HOUSE	1515	900	—	2415	600	84	684	$510.00	
S/T 2	WEST OF THE MOON	1470	—	887	2357	192	234	426	610.00	
S/T 3	SUN BUCKET	1256	600	—	1856	273	555	828	280.00	
S/T 4	SOLAR HOUSE ON THE PRAIRIE	1246	—	756	2002	*	333	333	685.00	*WATER PRE-HEAT SYSTEM
S/T 31	JACKSONVILLE	1915	—	—	1915	*	783	783	550.00	*WATER PRE-HEAT SYSTEM
S/T 6	WINGS	895	—	704	1599	*	630	630	375.00	*WATER PRE-HEAT SYSTEM
S/T 7	RADIO SOLAR MUSIC HALL	794	824	596	2214	518	216	734	460.00	
S/T 8	SUN BATH	1089	—	1075	2164	317	252	569	340.00	
S/T 32	EARTHWORM	1483	—	—	1483	*	210	210	540.00	*WATER PRE-HEAT SYSTEM
S/T 10	GREENHOUSE GALLERY	1166	—	1025	2191	*	680	680	780.00	*WATER PRE-HEAT SYSTEM
S/T 11	SOLAR TRI-LEVEL	815	195	980	1990	850	60	910	390.00	REVERSE PLAN - ADD $100.00
S/T 33	SPACEDOME I	1365	—	1090	2455	*	68	68	780.00	*WATER PRE-HEAT SYSTEM
S/T 13	SUN TREK	1260	—	578*	1838	220	—	220	390.00	*INCLUDES 112 SF LOW HEADROOM FINISHED SPACE
S/T 34	TAHOE	1022	—	1054	2176	*	908	—	550.00	*WATER PRE-HEAT SYSTEM
S/T 15	SKY BUBBLES	1066	—	886	1952	567	—	567	520.00	GARAGE ONLY $170.00 (PLEASE DIAGRAM ORIENTATION)
S/T 16	OLD GREEN THUMBS	900	952	543	2395	253	586	839	620.00	PLEASE SPECIFY BILLIARD ROOM OR EXTRA BEDROOMS
S/T 17	THE CONVECTION CONNECTION	1019	—	1014	2033	*	744	744	370.00	*WATER PRE-HEAT SYSTEM
S/T 18	INNERPEACE	1444	826	575	2845	351	240	591	530.00	PLEASE SPECIFY TYPE OF ROOF
S/T 19	BEHIND CLOSED DOORS	972	—	650	1622	*	350	350	380.00	*WATER PRE-HEAT SYSTEM
S/T 20	A TOUCH OF GREEN	1155	—	982	2137	468	184	652	540.00	REVERSE PLAN - ADD $100.00
S/T 21	EXPLORER	1452	798	482	2732	525	186	711	720.00	
S/T 35	DES MOINES	931	—	931	1862	*	576	576	550.00	*WATER PRE-HEAT SYSTEM
S/T 36	QUAIL'S NEST	1298	—	—	1298	*	226	226	390.00	*WATER PRE-HEAT SYSTEM
S/T 24	SUN CRUISER	822	896	448	2166	620	—	620	425.00	PLEASE SPECIFY "LOFT" OR "MASTER SUITE" VERSION
S/T 25	CLASSIC FUNK	1161	—	1065	2226	*	420	420	610.00	*WATER PRE-HEAT SYSTEM
S/T 26	THE VAGABOND'S CABIN	930	—	392	1322	135	105	240	310.00	
S/T 37	SPACEDOME II	1086	1358	—	2444	*	138	138	780.00	*WATER PRE-HEAT SYSTEM
S/T 28	AGE OF LUXURY	1272	1363	—	2635	—	600	600	680.00	
S/T 29	APOLLO	1030	—	997	2027	504	224	728	620.00	
S/T 30	XP 230-7	1535	—	873*	2408	324	576	900	720.00	*INCLUDES 120SF LOW HEADROOM FINISHED SPACE

Request for Plans

Space Time Designs, Inc.

P.O. BOX 1989, SEDONA, AZ 86336

Name ______________________________

Title ______________________________

Company ______________________________

Address ______________________________

City ______________ State ______________ Zip ______________

May we have your phone number? It will help us if we need to call you personally about your order.

Area Code ☐☐☐ Telephone Number ☐☐☐ ☐☐☐☐

Dear George,

☐ Please send the complete house plan package for the Space/Time Designs Solar home(s) indicated below.

I have enclosed a money order, or cashier's check, as part of the consideration for this order. I agree not to use these serialed and copyrighted plans more than once without written permission of Space/Time Designs, Inc. I understand that the drawings may not be reproduced in whole or part without the written consent of Space/Time Designs, Inc. I agree as additional consideration not to reproduce, loan to others or repeat building these designs without the written permission of Space/Time Designs, Inc. I understand that this purchase is subject to the limitations contained on the reverse side of this form. My building site address is:

____

Signature	Date

S/T Design Number	Title and Options (If Desired)	Complete Pkg @ ______	Planning Set @ ___ (15% of com pkg. price)	Total
Building Package without Plans — (See description pg. 23)			$45.00	

Total Amount of Order ☐

Request for Plans

Space Time Designs, Inc.

© Copyright Space/Time Designs, Inc. 1981

P.O. BOX 1989, SEDONA, AZ 86336

Name ______________________________

Title ______________________________

Company ______________________________

Address ______________________________

City ______________ State ______________ Zip ______________

May we have your phone number? It will help us if we need to call you personally about your order.

Area Code ☐☐☐ Telephone Number ☐☐☐ ☐☐☐☐

Dear George,

☐ Please send the complete house plan package for the Space/Time Designs Solar home(s) indicated below.

I have enclosed a money order, or cashier's check, as part of the consideration for this order. I agree not to use these serialed and copyrighted plans more than once without written permission of Space/Time Designs, Inc. I understand that the drawings may not be reproduced in whole or part without the written consent of Space/Time Designs, Inc. I agree as additional consideration not to reproduce, loan to others or repeat building these designs without the written permission of Space/Time Designs, Inc. I understand that this purchase is subject to the limitations contained on the reverse side of this form. My building site address is:

____

Signature	Date

S/T Design Number	Title and Options (If Desired)	Complete Pkg @ ______	Planning Set @ ___ (15% of com pkg. price)	Total
Building Package without Plans — (See description pg. 23)			$45.00	

Total Amount of Order ☐

Limitations

1. Space/Time Designs, Inc. represents that all plans offered are suitable and may be used to erect the structures which they describe.

2. Purchaser assumes sole responsibility for all aspects of the construction of the properties rendered in the plans contained in this book and hereby releases Space/Time Designs, Inc. and agrees to hold them harmless from any and all defects related to the construction of any such structure.

3. All designs shown have been prepared from accurately drawn working drawings. Square footage figures, where listed, are calculated from outside building dimensions and exclude greenhouse floor area, garage space, most mechanical rooms and active heat storage area. All square footage figures, greenhouse volumes, heat storage volumes, collector areas and passive apertures have been approximated as carefully as possible, are included solely for the convenience of the purchaser and may be useful in your evaluations; however, Space/Time Designs, Inc. makes no representation or warranty as to their absolute accuracy. Suggested passive and active mass storage areas are higher than would likely be used, but are shown in this book as total potential.

4. Space/Time Designs, Inc. believes these designs to be effective and solar-efficient for the right family and climate. However, solar efficiency may vary depending on many factors such as local climate. Therefore, Space/Time Designs, Inc. makes no representation or warranty as to the degree of solar efficiency of these designs.

All these designs incorporate universally standard wood frame building systems but some details may be more indigenous to those used in the Pacific Northwest. Roof designs are based on 40 lbs. per square foot live and dead loads. Floor loads are based on 50 lbs. per square foot live and dead loads. Wind, seismic and snow loads should be carefully evaluated by local professionals.

Limitations

1. Space/Time Designs, Inc. represents that all plans offered are suitable and may be used to erect the structures which they describe.

2. Purchaser assumes sole responsibility for all aspects of the construction of the properties rendered in the plans contained in this book and hereby releases Space/Time Designs, Inc. and agrees to hold them harmless from any and all defects related to the construction of any such structure.

3. All designs shown have been prepared from accurately drawn working drawings. Square footage figures, where listed, are calculated from outside building dimensions and exclude greenhouse floor area, garage space, most mechanical rooms and active heat storage area. All square footage figures, greenhouse volumes, heat storage volumes, collector areas and passive apertures have been approximated as carefully as possible, are included solely for the convenience of the purchaser and may be useful in your evaluations; however, Space/Time Designs, Inc. makes no representation or warranty as to their absolute accuracy. Suggested passive and active mass storage areas are higher than would likely be used, but are shown in this book as total potential.

4. Space/Time Designs, Inc. believes these designs to be effective and solar-efficient for the right family and climate. However, solar efficiency may vary depending on many factors such as local climate. Therefore, Space/Time Designs, Inc. makes no representation or warranty as to the degree of solar efficiency of these designs.

All these designs incorporate universally standard wood frame building systems but some details may be more indigenous to those used in the Pacific Northwest. Roof designs are based on 40 lbs. per square foot live and dead loads. Floor loads are based on 50 lbs. per square foot live and dead loads. Wind, seismic and snow loads should be carefully evaluated by local professionals.